METEOROLOGY: PREDICTING THE WEATHER

INNOVATORS

Astronomy: Looking at the Stars
Aviation: Reaching for the Sky
Communications: Sending the Message
Computers: Processing the Data
Construction: Building the Impossible
Diseases: Finding the Cure
Forensics: Solving the Crime
Genetics: Unlocking the Secrets of Life
Medical Technology: Inventing the Instruments
Meteorology: Predicting the Weather
Motion Pictures: Making Cinema Magic
Space Flight: Crossing the Last Frontier
Weapons: Designing the Tools of War

METEOROLOGY: PREDICTING THE WEATHER

Susan and Steven Wills

The Oliver Press, Inc.
Minneapolis

This book is dedicated to my mother, Peg Barnes, a lover of sunrises, sunsets, and skies filled with clouds.—Susan
This book is also dedicated to the first meteorologist I ever knew—my dad—who taught me how to read the skies and the wind.—Steven

The Oliver Press, Inc.
Charlotte Square
5707 West 36th Street
Minneapolis, MN 55416-2510

Library of Congress Cataloging-in-Publication Data
Wills, Susan, 1948-
Meteorology: predicting the weather / Susan and Steven Wills.
p. cm. — (Innovators ; 12)
Includes bibliographical references and index.
Contents: Daniel Gabriel Fahrenheit and the standard temperature scale — Benjamin Franklin and the electricity of lightning — Luke Howard and cloud classification — Vilhelm Bjerknes and fronts and pressure systems — Carl-Gustaf Rossby and planetary waves — Jule Charney and computer forecasting — Howard Bluestein and storm chasing.
ISBN 1-881508-61-7
1. Meteorologists—Biography—Juvenile literature. 2. Meteorological instruments—Juvenile literature. 3. Weather forecasting—Technique—Juvenile literature. [1. Meteorologists. 2. Weather forecasting. 3. Meteorological instruments.] I. Wills, Steven R. II. Title. III. Series.

QC858.A2W55 2004
551.5'092'2—dc21
[B]

2002041031
CIP
AC

ISBN 1-881508-61-7
Printed in the United States of America
10 09 08 07 06 05 04 8 7 6 5 4 3 2 1

CONTENTS

INTRODUCTION

From Myth to Measurement

In Greek mythology, Zeus was the most powerful of the gods, but he was not the most worshiped. That honor fell to Demeter, goddess of the harvest, because the success or failure of the harvest was a life-and-death issue for the ancient Greeks. Demeter had one daughter, Persephone, the maiden of spring. One day, Persephone attracted the attention of Hades, god of the underworld. Hades was immediately taken by her beauty, and Persephone was immediately taken by Hades—who forced her to be his bride and live with him in the underworld.

Demeter searched for her daughter and finally found out that she had been kidnapped by Hades. Deeply depressed, Demeter would not allow any seeds to sprout or any fruit to grow in the land. This brought Zeus into the conflict. (After all, without any harvest, people would soon die, and then who would worship the gods?) In the end, Zeus struck a deal between Demeter and Hades. For half the year,

Throughout history, human beings have been at the mercy of the weather, which can bring not only life-giving sun and rain, but also destructive storms like the tornado shown here.

Persephone would stay with Hades in the underworld, and for the other half she would return to stay with Demeter. Demeter wasn't very happy with the arrangement, but she agreed. During the time Persephone was in the underworld, however, Demeter would not allow any food to grow, and she covered the world with ice and snow.

This well-known Greek myth not only served to explain the changing of the seasons, but it was also a reminder that, although winter might be bitterly cold and summer heat might be unbearable, no season lasts forever. The only thing that never changes about the weather is that it always changes.

Threatening life and property, ocean water floods ashore during a hurricane (a system of severe rotating winds that forms over the ocean). In one of the worst natural disasters in U.S. history, a 20-foot wave caused by a hurricane destroyed 3,000 homes and killed 6,000 people in Galveston, Texas, in 1900. Weather forecasting was not advanced enough at the time to give adequate warning of the storm.

Lightning (shown here over Boston in the 1960s) has always been a dramatic and unpredictable force. Striking about 40 million times a year in the United States alone, it is the nation's second most fatal type of weather (flooding is the first).

Throughout recorded history, sudden and seemingly random destruction caused by the weather has reminded people of their connection to the earth and their helplessness when it seems to turn on them. Major weather events—a cold winter that ruined the crops of medieval Europe, a typhoon and accompanying flood that brought death and homelessness to poverty-stricken Bangladesh in the 1970s, or wildfires that sweept across rain-starved Colorado in 2002—reveal humanity's need to understand the forces that shape our planet and our lives.

Weather is a fact of life. All human beings experience it from their first breath to their last. It has made a difference in battles, love affairs, and sports. But more than with most things, we take its daily variations for granted. It takes weather anomalies—blizzards, tornadoes, hurricanes, or temperature extremes—to remind us that the world is still governed by nature.
—Francesca Lyman, *The Greenhouse Trap*

The Greeks were not the only early civilization to worship gods associated with the weather. Ancient Egypt's two primary gods, Ra and Osiris, brought rain and controlled the Nile River (the center of Egyptian agriculture). Babylonians worshiped Marduk, the god of the atmosphere; early civilizations of India worshiped Indra, the god of rain and storms; and some Native American myths include stories of the Thunderbird, who brought lightning with the winking of his eyes and thunder from the flapping of his wings. Weather gods were central to all cultures. After all, people could hunt wild beasts or battle invaders, but no one could escape the weather. If the rains stopped and the crops withered, or if a devastating storm ripped through fields and villages, they simply had to stand by and suffer the consequences, powerless to do anything about it. It was easy to view natural disasters as punishments aimed at disobedient people who displeased the gods.

Despite this, humans still tried their hardest to understand and predict the weather. Many early civilizations observed and kept records of weather patterns, and some went even further. The historical record shows that it may have been the ancient Chinese who were the first to scientifically study the weather. For example, they developed a way to measure humidity by weighing a piece of dry charcoal and then exposing it to the air, where it would absorb moisture. When the charcoal was weighed a second time, its increase in weight indicated the amount of moisture in the air—and thus the likelihood of rain.

humidity: the amount of water vapor in the air

Much later, the Greeks were probably the first to contemplate and discuss the actual causes of weather. They even gave this effort a name: meteorology, or the study of meteors. For them, the word "meteor" described anything that appeared in the space between Earth and the Moon. Besides meteors, this could include clouds, rain, snow, fog, hail, thunder, and rainbows.

Thales of Miletus (620-555 B.C.) is recognized as the first Greek meteorologist. He attempted to understand the hydrologic (water) cycle that creates rain and snow, and although he wasn't clear on evaporation, his view was fairly accurate. Anaximander (611-547 B.C.) was the first to describe wind as a flow of air. Working with Anaximenes (585-528 B.C.), he also theorized that thunder was the result of air smashing its way through clouds—a violent passage that sparked lightning. Anaxagoras (499-427 B.C.) believed that lightning occurred when fire from the "aether" high above descended to Earth, and thunder was the sound of this fire hissing through the moisture of the clouds. More correctly, he stated that sea breezes were formed when warm air rose above colder air. Although these early scientists rarely got their details exactly right, they were all notable in viewing weather as something explainable, rather than as the wrath of angry gods.

Empedocles (492-430 B.C.) gathered together all the known theories of meteorology and concluded that the universe was made up of four elements: earth (heat), air (cold), fire (dryness), and water (moisture). The interaction between these elements

hydrologic cycle: the continuous circulation of water through the biosphere. Ocean water is heated by the sun until it evaporates and is lifted into the atmosphere by rising air. There, the vapor cools and condenses into clouds, which can produce moisture in the form of precipitation (rain or snow) that falls to Earth and is absorbed into the ground or runs off through streams and rivers to the ocean. This water eventually evaporates into the atmosphere, and the cycle begins anew.

Thales is also said to have issued the first seasonal crop forecast. Based on past olive harvests, he predicted a bumper crop for the following year. Supposedly, he was so confident in his forecast that he reserved all of the olive presses in the area ahead of time, then made a profit by leasing them to farmers when the large harvest arrived as he expected.

produced weather. For instance, heat and moisture created clouds and rain, while dryness made thunder and wind. This idea was adopted by Aristotle (384-322 B.C.), who is usually given credit for the theory. The four elements were at the heart of Aristotle's *Meteorologica*, which became the recognized universal text on the weather for the next 2,000 years. People in the centuries to come simply reacted to Aristotle's statements without questioning the ideas behind them. Even Roger Bacon (A.D. 1214-1294), one of the great thinkers of the Middle Ages, agreed with Aristotle. Bacon did, however, insist on experimental support of these ideas, and this was the first hint that the scientific world would be ready to break away from Aristotle—eventually.

Aristotle wrote about nearly every subject, including logic, ethics, politics, poetry, biology, and physics. Most of his published writings have been lost, and his ideas are mainly known through student lecture notes and textbooks from the Lyceum, the school he founded in Athens.

By the 1600s, many people were looking to astrologers to predict the weather. The astrologers based their predictions on two ideas: that the Moon and the planets controlled the weather, and that animals were more sensitive to changes in the weather than humans were. They compiled and consulted long lists of signs that would foretell fair or foul weather. Fair weather, for example, might be indicated by cranes flying out to sea or oxen lying on their left sides. If ants were seen carrying their eggs from their nests, it meant rain, and this could be confirmed if dogs scratched at the ground. If those dogs were seen rolling on the ground, or if the oxen were lying on their right sides, look out—a storm was on its way.

This method of weather prediction was based on a simple premise: events that had been seen to precede certain weather conditions on previous occasions could be used to forecast future ones. The reasoning was correct to a certain extent, and some early weather wisdom contains a grain of truth. For example, the phrase "Red sky at night, sailor's delight; red sky at morning, sailors take warning" came from observations that brightly colored sunsets seemed to indicate clear weather. Scientists now know that colorful sunsets are caused by dust particles in the air reflecting the sun's light, so they are more likely to happen in dry conditions. Since most weather comes from the west, a red sunset could mean that drier air was on its way. A red sunrise, on the other hand, could mean that the dry air had passed to the east and that wetter air—and storms—

astrology: the study of the positions of the Sun, Moon, stars, and planets in the belief that they influence earthly events and human fortunes

The astrologers may have been on the right track, at least partly. It is likely that animals and plants are able to perceive subtle changes in the atmosphere that affect the weather. Some modern meteorologists study the coats of woolly caterpillars, the nests of squirrels, or the migration of birds for insight into the coming season. Although there have been no certain connections between animal behaviors and long-range weather patterns, the research is ongoing.

Although most early beliefs about weather were disproved by science, remnants still remain in our language. For instance, we say dew "falls" rather than condenses on the grass, and lightning "strikes" instead of being conducted as electricity. Moreover, "the elements" remains a common term for the weather.

were coming. Other pieces of weather folklore, however, were simply wrong—like the saying "Lightning never strikes in the same place twice." (The Empire State Building is struck about 23 times every year!)

For truly accurate forecasting, a more thorough understanding of the weather was needed. When René Descartes (1596-1650) developed the basis for the modern scientific method, meteorology finally began to move away from the theories of both Aristotle and the astrologers. Descartes suggested scientists should draw their conclusions by making observations, measuring and recording data, examining the data to form a general hypothesis, and then applying that hypothesis to see if it held true. In this process, new ideas would not simply be accepted as facts, but instead would be questioned and tested to make sure they worked.

The scientific method, however, depended upon measurement—and there were no reliable instruments to measure things related to meteorology. Accurate ways of gathering information were needed before Benjamin Franklin could notice the changes in ocean temperatures that led him to map the current known as the Gulf Stream. Luke Howard's observations of clouds had to be combined with daily measurements of weather conditions to form his cloud identification system. Readings from the ground and from aloft were necessary for Vilhelm Bjerknes and his student Carl-Gustaf Rossby to determine how large air currents affected weather—and how weather prediction could help their respective nations during times of war. Jule

Charney required numerical data to program the computers that would revolutionize forecasting and lay the groundwork for the the modern age of meteorology. Technology also allowed contemporary pioneers such as Howard Bluestein the opportunity to peer into the heart of weather's most devastatingly beautiful storms.

None of these efforts to expand the boundaries of science would have been possible without accurate instruments—one of the most basic of which was the thermometer. Galileo Galilei (1564-1642) built a primitive thermometer in 1607, but he called it a "little joke." It could show changes in temperature, but it was not marked off in meaningful units of measurement and could not produce results that could be duplicated. Science demanded reliable thermometers based on a standard system of measurement. Over the next century, Galileo's device intrigued an impressive crowd of world-renowned thinkers who developed and improved it. But the innovator who finally transformed a "joke" into the most central tool of meteorology was simply a roving instrument maker named Daniel Gabriel Fahrenheit. With him began the scientific measurement of the weather.

A physicist, mathematician, and astronomer, Galileo Galilei is best known for building one of the earliest telescopes. His discoveries with the instrument demonstrated that the Sun (rather than Earth) was the center of the universe.

CHAPTER ONE

Daniel Gabriel Fahrenheit and the Standard Temperature Scale

The first thing people usually want to know about the day's weather is: How hot or cold will it be? It seems like a simple question, but if it were as simple as it seemed, no one today would recognize the word "Fahrenheit."

Temperature is such a basic measurement that we often take it for granted. Simply stated, the temperature of something is how hot or cold it feels. Scientists don't refer to the "coldness" of something, however. Cold is only a lack of heat. Ice does not feel cold because it has a quality called "coldness," but because our hands have more heat than the ice. Thus, temperature is essentially how much heat an object emits. Even more scientifically, temperature is a measurement of the kinetic energy (energy due to motion) of an object's atoms as that energy is transferred to another object in the form of heat, following the laws of thermodynamics. When you stick your finger into a glass of water in which the atoms are

Daniel Gabriel Fahrenheit (1686-1736) not only built the most accurate thermometers of his time, but also developed the first practical scale for measuring temperature.

thermodynamics: the branch of physics that studies the transformation of heat into other forms of energy. There are three laws of thermodynamics: (1) Energy cannot be created or destroyed, but can be transformed into other forms; (2) Heat will not pass from a colder body to a hotter one; and (3) It is impossible to reduce anything to absolute zero (the complete absence of heat).

moving relatively slowly, you don't feel much heat. In fact, if the atoms of your finger are moving faster than the water atoms, then the water will feel cold. But if you heat that glass of water on the stove, in a couple of minutes the water atoms will be moving so fast you will not want to stick your finger into them. They will soon begin to move so fast that they will boil, turning from a liquid into a gas.

The fact that heat is transferred so easily from one object to another (from the water to your finger, for example) means that it is measurable. But there are two obstacles to measuring heat. The simpler problem is to invent a device that can record heat. The more difficult problem (although it seems simple at first glance) is to devise a standard system of measurement that allows the temperature measured in London, for instance, to be compared to another measurement taken in Rome.

The answer to the first problem was to invent a thermometer. A thermometer is an instrument that measures temperature quantitatively—in a way that can be set to some number scale. To do this, scientists needed a device containing a substance that would change in a predictable manner as it was heated. The earliest such instruments, created in the early 1600s, were called thermoscopes. They relied on the effect of temperature on the air inside a glass tube that was open on one end. The open end was placed in a pan of colored liquid (such as water or wine), and as the air inside the tube expanded due to heat or contracted due to cold, the liquid would fall or rise in the tube. Early thermoscopes were

A thermoscope in 1688

highly decorative, etched with ornate designs and mounted on carved wooden panels, but they were also fragile and impractical. They could not be used to find the actual temperature—only whether the temperature was rising or falling.

In 1641, the first sealed thermometer was made for Grand Duke Ferdinand II of Tuscany (1616-1670). Known as a spirit thermometer, it relied on the effect of temperature on colored alcohol within a closed glass tube. There were degree markings

Ferdinand II initiated the world's first major weather monitoring project in 1654. He established about a dozen observation stations in northern Italy and across central Europe, each equipped with the same instruments and a standardized procedure for recording data. The program continued for 13 years.

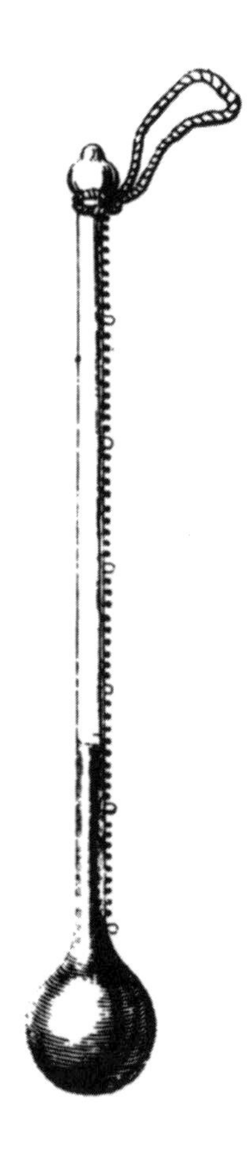

along the tube, but they weren't very useful, since they were not calibrated (set) to any standard scale. Three years later, however, Robert Hooke (1635-1703) took the spirit thermometer to the next step by establishing a scale of measurement based on a fixed point—the temperature at which water froze. His design was used by England's Royal Society of scientists until 1709.

Calibrating an accurate scale, however, requires two fixed points, not one. Even if everyone agrees that the freezing point of water is "0," how far up the scale is "10," and what does "10" mean? It was not until 1702 that Danish astronomer Ole Rømer (1644-1710) began building thermometers with a

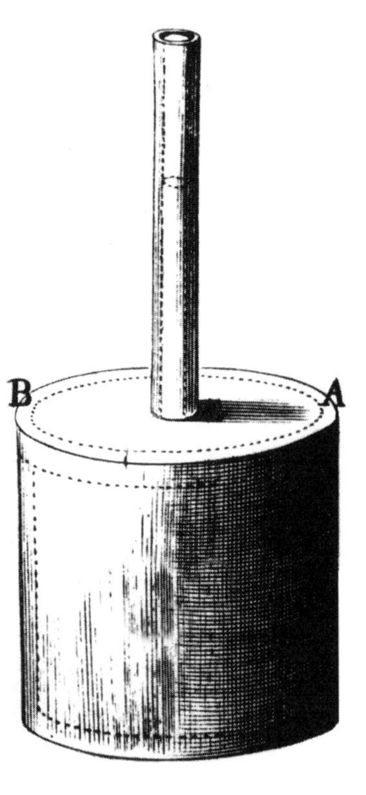

Above: a spirit thermometer with 50 degree markings. At right: Hooke's thermometer. He established its measuring scale by placing the glass tube full of wine into a metal cylinder, also filled with wine. When they were heated, he could compare the expansion of liquid in both to create a more standardized scale.

Ole Rømer built many instruments besides thermometers. He is shown here using his innovative transit telescope, which he created in the 1680s to measure the position of stars as they crossed a certain point over Earth. Rømer's other scientific contributions included the first calculation of the speed of light, which he found to be 140,000 miles per second (modern measurements give the speed as 186,282 mps).

scale based on two points: the freezing point of water at one end and the boiling point of water at the other. This was a good beginning, but Rømer's thermometers were still not up to the task of setting the world's standard for accurate measurement of temperature. For that, Rømer would have to pass his expertise to someone younger who could move the science forward. Fortunately, two years before his death, he was visited by a 22-year-old instrument maker named Fahrenheit.

Rømer successfully used his thermometer to record the daily temperatures in Copenhagen during the exceptionally cold winter of 1708-1709.

Daniel Gabriel (sometimes called Gabriel Daniel) Fahrenheit was born May 24, 1686, in

Danzig (now Gdansk), Poland. His father was a wealthy merchant whose family had moved to Danzig a generation before. Daniel was the oldest of six children, with three younger sisters and two younger brothers. Although the family enjoyed an adequate income, he received no formal education. When he was only 15 years old, both of his parents died suddenly, probably from eating poisonous mushrooms. Daniel's guardian sent him to Amsterdam in the Netherlands to become an apprentice and learn about business, while his siblings were placed in foster homes.

Apparently, Daniel Fahrenheit was not really interested in the life of a merchant, and, in fact, he seems to have gotten into some trouble regarding his apprenticeship. In 1707, probably to escape from punishment as much as to increase his education, Fahrenheit began to travel around Europe. He found himself wandering into the company of scientists in cities such as Berlin, Leipzig, and Dresden. In 1708, his budding interest in making scientific instruments brought him to Copenhagen and the company of Ole Rømer. He formed a close friendship with the astronomer, learning from him the craft of making accurate thermometers. It was Fahrenheit's plan to return to Amsterdam and enter the thermometer business—but first, he needed to make some improvements to Rømer's designs.

THE BREAKTHROUGH

Although the technology of spirit thermometers was satisfactory, individual instruments varied depending on the skill of the manufacturer. Many factors made comparisons between different thermometers inconsistent. Because alcohol did not expand at an even rate, spirit thermometers could not be calibrated in even increments. Alcohol did not freeze easily, but it had a low boiling point, so it could not be used for high temperatures because it would begin to change into a gas. Using water in thermometers was no better. It could not be used for temperatures near its boiling or freezing points, since the water would begin to change at those points. Water was also unpredictable, expanding differently at different altitudes.

In 1714, Fahrenheit developed a more accurate thermometer by substituting mercury for alcohol or water. Mercury had several advantages: it had a lower freezing point and a higher boiling point, so it remained a liquid within a wider temperature range. Mercury also expanded and contracted at a very constant rate, even at higher elevations. For this reason, the dividing lines between degrees could be marked at regular and predictable intervals on the instrument. In addition, mercury's bright, silvery appearance made the thermometer easy to read. Fahrenheit invented a method for cleaning the mercury to prevent it from sticking to the inside of the tube, giving it another advantage over colored water or alcohol, which tended to muck up the glass. After studying

mercury: a heavy, silver-white metallic element

how glass expanded at high temperatures, he even designed his tube to eliminate this effect, making the instrument still more reliable.

Fahrenheit's next step was to decide on the key points for his measuring scale. Rømer had used the freezing and boiling points of water, which he set at 7.5° and 60°, respectively. Rømer knew, however, that the top part of this scale would be useless for studying the weather, since air temperatures rarely (if ever) approached the boiling point of water. Thus, he also used a third fixed point, human body temperature, which he called "blood heat" and set at 22.5°. Rømer built several thermometers with 22.5° as the upper limit instead of 60°. Since fewer degree markings would need to fit on the tubes, they could be spaced farther apart and the differences between them would be much more visible—and more useful for meteorology.

Fahrenheit agreed with this strategy, but he believed Rømer's scale was inconvenient because it involved so many fractions. In addition, anyone who had sailed the North Atlantic Ocean in winter knew that ocean water could remain liquid even below the temperature at which pure water would freeze. Fahrenheit decided to add salt to his water in order to find out its lowest possible freezing point. The temperature at which salt water froze became Fahrenheit's 0°. This proved useful, allowing him to measure air temperature in the middle of winter without such frequent use of negative numbers. Fahrenheit's second fixed point was the freezing point of fresh water, which he set at 30°. The third was blood heat, for which he settled on 90°. Using this scale, Fahrenheit then measured the boiling point of water as 212°.

Fahrenheit now had the skill to build uniform glass instruments, a reliable liquid for measurement, and a logical scale with several fixed points. He returned to the Netherlands, and in 1717 he began to make and sell his own thermometers. His thermometers were the first that looked and operated like predictable, scientific instruments rather than decorations. The Fahrenheit scale was also the first to gain wide acceptance. Earlier thermometers had often been mounted on large wooden plaques to accommodate as many as 12 different measuring scales. The need for a universal scale had become apparent as scientists began to share information over long distances. Certainly, any study of meteorology would require accurate comparisons among varied locales and climates. Once the scientific community accepted the Fahrenheit scale as its standard, scientists could begin to write about temperature in a way that would be meaningful to colleagues.

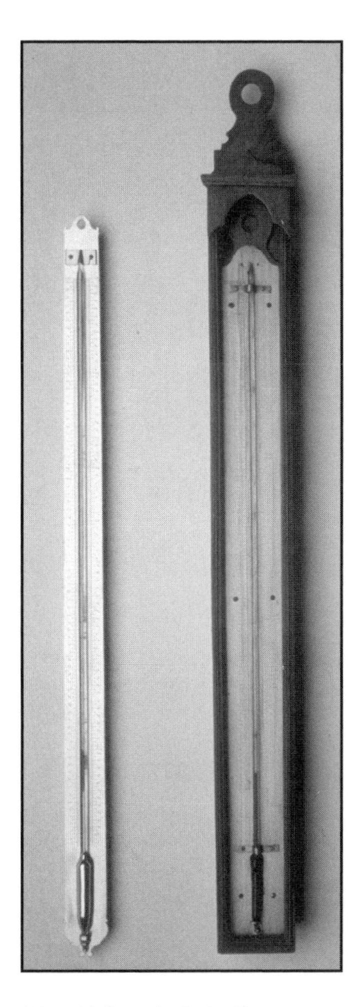

Two Fahrenheit-built thermometers

Fahrenheit adjusted his freezing point of fresh water from 30° to 32° so that the interval between the freezing and boiling points would be exactly 180°. He also reset blood heat at 96°. After Fahrenheit's death, however, blood heat was eliminated as a fixed point on the scale because it was not very predictable. Instead, the boiling point of water (212°) became the upper point. The result was the temperature scale that is known today as degrees Fahrenheit (°F).

Standard body temperature on the Fahrenheit scale is now set at 98.6°.

atmospheric pressure: pressure caused by the weight of the atmosphere (also called barometric pressure)

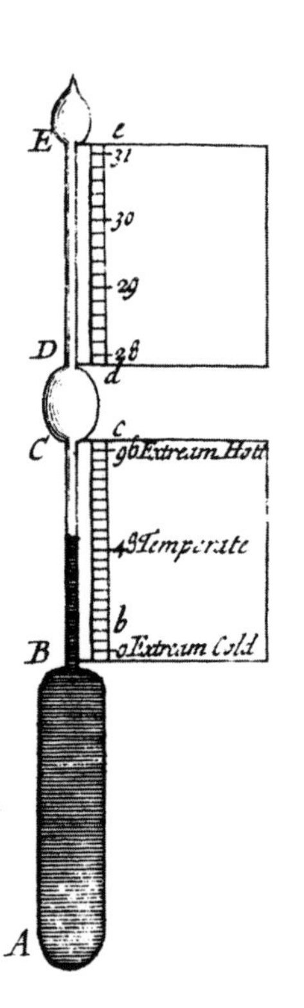

Fahrenheit's hypsometric thermometer. The top part was marked off in barometric units (in this case, inches of mercury) so that when it was placed in boiling water, it would show the atmospheric pressure.

THE RESULT

The Fahrenheit scale was soon adopted as the official measure of temperature in the Netherlands and Great Britain. Even though Daniel Fahrenheit was not a trained scientist, his expertise at instrument making earned him an election to the prestigious Royal Society in 1724. And even though he was never much of a writer, in that year he published five papers in the society's magazine, *Philosophical Transactions*. In one of them, he described the boiling points of several liquids, and in another he reviewed his decision-making process in the development of the Fahrenheit scale.

Although he is remembered mostly for his precise thermometers, Fahrenheit left his mark in other areas as well. Noting that the boiling points of various liquids varied with atmospheric pressure, he constructed a hypsometric thermometer—a device that could be used to measure the boiling point of water and then determine the atmospheric pressure. Fahrenheit also invented an early hydrometer, used to measure the density of liquids by comparing them to the density of water. He even discovered that fresh water could remain in a liquid state below its freezing point. In 1736, he patented a pump that could be used to drain water from the Dutch lowlands. On September 16 of that year, at the age of 50, Daniel Gabriel Fahrenheit died at The Hague, Netherlands. A skilled colleague, Hendrik Prins, continued the manufacture of his thermometers.

A modern thermometer using the Fahrenheit scale. In addition to labeling the "blood heat" point, this instrument also marked weather-related points such as "temperate" and "summer heat." On this cold day in 1938, however, the air temperature was below "freezing."

In 1742, an astronomer named Anders Celsius (1701-1744) established a scale with 100° as the freezing point of water and 0° as the boiling point. He felt his scale would be more practical, since it could be used in the winter without relying on any negative numbers. Then, three years later, Swedish scientist Carolus Linnaeus (1707-1778) flipped the Celsius scale over, establishing the Centigrade (meaning "100 divisions") scale with 0° as the freezing point and 100° as the boiling point. In the late 1800s, a variation of the Centigrade scale using a gas mixture in place of mercury was adopted for the

Anders Celsius spent his career teaching and working in his birthplace, Uppsala, Sweden. An observatory was built there in his honor in 1740, and he served as its first director.

The Centigrade scale was renamed because surveyors and architects often use a metric measurement of angles called a "grade." One one-hundredth of this angle would be a centigrade, which could be confused with the temperature scale.

international measurement of temperature. The terms "Celsius" and "Centigrade" began to be used interchangeably, so to avoid confusion, the Ninth General Conference of Weights and Measures declared in 1948 that the Centigrade scale would officially be called "Celsius" (°C). By 2003, the United States and a few smaller countries were the only nations that still used the Fahrenheit standard instead of the Celsius scale.

The modern scientific community uses yet another scale. The thermodynamic temperature scale, known as the Kelvin scale (K), was developed by William Thompson, the first Baron Kelvin (1824-1907), in 1848; the International Committee of Weights and Measures adopted it in 1933. Its units are the same size as those of the Celsius scale—so that a change of 10°C is the same as a change of 10K—but its fixed points are different. One fixed point on the Kelvin scale is the triple point of water, the temperature at which water, water vapor, and ice can coexist. It is set at 273.16K, equal to 0.01°C. The other is "absolute zero," the point at which molecular motion stops, which is set at 0K (-273.15°C). Because this is an impossibly cold temperature, using the Kelvin scale never requires negative numbers.

The measurement of temperature has clearly become more sophisticated since Fahrenheit's development of a reliable instrument and a standard scale. Although mercury thermometers still exist, many modern thermometers are electronic. Scientists can use thermometers based on the expansion of various gases or metals, or devices such as radiometric thermometers, which measure temperature by reading the radiation an object emits. Nevertheless, Fahrenheit was the first to make it possible for someone working in one part of the world to describe temperatures to someone elsewhere. As a matter of fact, on his way across the Atlantic Ocean in 1775, American diplomat Benjamin Franklin used a thermometer to note that one area of the ocean was warmer than the surrounding water. He recorded his findings using the Fahrenheit scale, which reliably communicated to his associates in France his discovery of the Gulf Stream.

CHAPTER TWO

Benjamin Franklin and the Electricity of Lightning

He was a printer, a publisher, a scientist, a philosopher, an inventor, a politician, a diplomat, an avid writer, and probably the best-known American of his day. His was the ultimate success story—the model for many rags-to-riches tales since—and, had he been a few years younger, he would probably have been the first president of the United States.

Such honors would not have been expected when Benjamin was born on January 17, 1706, to Josiah Franklin and his second wife, Abiah Folger. He was the 15th of 17 children and the youngest of 10 boys in this working-class Boston family. His father, who made candles and soap, believed that his sons should be put to meaningful work as soon as possible. As a result, Benjamin was given just two years of formal education.

Although he is better known as one of the Founding Fathers of the United States, Benjamin Franklin (1706-1790) also made important advances in the science of meteorology.

Life, however, provides its own form of schooling, and in this, Benjamin was a star pupil. Josiah

Franklin noticed that his youngest son was constantly reading books. After Benjamin had worked two years in the family candle business, his father decided that he would be better suited to the job of a printer. So at the age of 12, Benjamin was sent to work as an apprentice for his older brother James, who was establishing his own printing shop. Josiah required Benjamin to sign an indenture (a legal contract binding someone into the service of another) promising to work for James for the next nine years.

Apprentices were not usually treated well. Add to this the growing envy James felt for his talented younger brother, and the result was a critical and often abusive relationship. Benjamin quickly learned the printing trade and yearned to strike out on his own. When government censors ordered James to quit his newspaper after he published articles criticizing the British colonial governor, James decided to print the paper in Benjamin's name instead. To do so, he had to free Benjamin from his indenture. Benjamin seized the opportunity and ran away.

The story of how Benjamin Franklin left Boston (at night and in secret) and traveled first to New York and finally to Philadelphia is one of the most frequently read sections of his famous autobiography. With little more than a change of clothes, a couple of dollars, and a knowledge of the printing trade, the 17-year-old arrived alone in Philadelphia in October 1723. He purchased a few penny loaves of bread and walked the length of the young city, finally falling asleep in a Quaker meeting-house. It was a small start for what was to become a large life.

Franklin began working for a local printer, and by the time he was 18, he had already become successful. The Pennsylvania governor even offered to subsidize him in his own printing business, and sent him off to London to purchase a press and supplies. Franklin soon found, however, that the governor's promises did not translate into any money, and for the next two years, he stayed in London working and saving the money for his press. In 1726, he returned to Philadelphia and set up a printing shop

Benjamin Franklin (center) as a young printer. Working a printing press required great upper-body strength, which Franklin built and maintained through one of his favorite leisure activities, swimming.

with a partner. They began publication of a newspaper, *The Pennsylvania Gazette*, in 1729. In 1733, Franklin printed the first issue of an annual journal that would establish his fame and his fortune—*Poor Richard: An Almanack*.

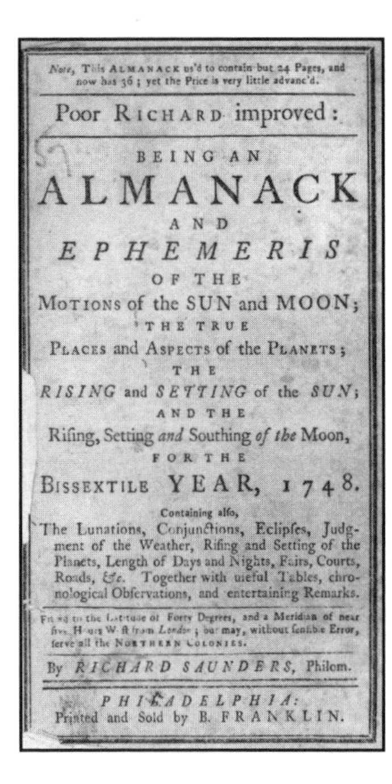

Note, This ALMANACK us'd to contain but 24 Pages, and now has 36; yet the Price is very little advanc'd.

Poor RICHARD improved:

BEING AN

ALMANACK

AND

EPHEMERIS

OF THE

MOTIONS of the SUN and MOON;

THE TRUE

PLACES and ASPECTS of the PLANETS;

THE

RISING and *SETTING* of the *SUN*;

AND THE

Riſing, Setting *and* Southing *of the* Moon,

FOR THE

BISSEXTILE YEAR, 1748.

Containing alſo,

The Lunations, Conjunctions, Eclipſes, Judgment of the Weather, Riſing and Setting of the Planets, Length of Days and Nights, Fairs, Courts, Roads, *&c.* Together with uſeful Tables, chronological Obſervations, and entertaining Remarks.

Fitted to the Latitude of Forty Degrees, and a Meridian of near five Hours West from *London*; but may, without ſenſible Error, ſerve all the NORTHERN COLONIES.

By *RICHARD SAUNDERS*, Philom.

PHILADELPHIA:
Printed and Sold by B. FRANKLIN.

The title page of the 1748 edition of Poor Richard's Almanack

Almanacs were calendars that contained information about weather, phases of the moon, eclipses, and tides, as well as horoscopes and medical advice. People loved reading almanacs, and printers loved printing them—after all, they were cheap and they could be printed whenever the press wasn't being used for something else. Printers usually hired astrologers to write their almanacs, and therefore they had to share the profits. Franklin, however, realized that most people read almanacs for amusement rather than accurate predictions, so he decided that he would serve as his own astrologer. Instead of presenting serious weather forecasts, he made fun of them through an imaginary character, Richard Saunders, otherwise known as Poor Richard. For example, here's what Poor Richard predicted for a future eclipse: "During the first visible eclipse Saturn is retrograde [seems to be moving backward]: For which reason the crabs will go sidelong and the rope makers backward. The belly will wag before and the A— [ass] shall sit down first."

In addition to weather forecasts, *Poor Richard's Almanack* was filled with poetry, trivia, humor, stories, and advice—all presented with a smile. Poor Richard became most famous for his sayings. Some have become part of American lore ("A penny saved is a penny earned"), some are full of wisdom ("Some are weather-wise: some are otherwise"), and many shine with humor ("Fish and visitors smell in three days"). *Poor Richard's Almanack* sold an average of 10,000 copies annually for the next quarter-century, and it remained so popular that Franklin's financial comfort was guaranteed for the rest of his life.

By age 30, Benjamin Franklin was already a local celebrity, and his energy and creative ideas continued to bring him fame and respect. He served as clerk of the Pennsylvania Assembly from 1736 until 1752, when he was elected as a member. From 1737 to 1753, he was the postmaster of Philadelphia, and in 1754 he became deputy postmaster-general for all the North American colonies. Franklin organized the first lending library in 1731 and the first volunteer fire company in 1736, and he also helped found the Pennsylvania Hospital and the College and Academy of Philadelphia (which later became the University of Pennsylvania) in 1749.

Comfortably wealthy, respected, important, and secure, Franklin was able to quit the day-to-day operation of his printing business and devote his time to other interests, especially science. As usual, Franklin began a pursuit by founding something—in this case, the American Philosophical Society in 1743. This group was organized to discuss, as he put it, "all philosophical experiments that let light into the nature of things, tend to increase the power of man over matter, and multiply the conveniences or pleasures of life." Franklin's gift was his ability to see science all around him. For example, on October 21, 1743, he planned an observation of a lunar eclipse in

Franklin's position as postmaster was a great advantage to his printing business. He had first access to the news from outside Philadelphia, and he could circulate his own newspapers by mail while stopping delivery from competing printers.

Benjamin Franklin had a son, William, with an unknown woman in about 1730. At around the same time, he began a relationship with Deborah Read, who had been abandoned by her husband. Since she was never officially divorced, she and Franklin did not legally marry, but they lived as husband and wife until her death in 1774. In addition to raising William, the couple had a son, Francis Folger (who died of smallpox at age four), and a daughter, Sarah.

Philadelphia. The Philadelphia sky was blotted with the clouds of an approaching storm from the northeast. One of Franklin's brothers in Boston, however, was able to view the eclipse. Franklin later found out that the storm over Philadelphia hit Boston about four hours later that night.

Immediately, Franklin went to work collecting data. He scoured newspapers and talked to travelers, merchants, and sailors looking for information about storms. He even tried following storms in person, on horseback, as they moved over the land. After charting the times and locations of storms from New England to Florida, Franklin concluded that they originated in Florida or Georgia and moved northeast at about 100 miles per hour. Even though the surface winds blew in off the ocean from the east or northeast, East Coast storms moved in a vast circular pattern from the southwest to the northeast, which Franklin correctly theorized was caused by warm air from the Gulf of Mexico meeting colder northern air. This discovery was just the first in a series of contributions to meteorology that would help improve weather forecasting and earn Franklin recognition in the scientific community.

To produce our North-East storms, I suppose some great heat and rarefaction of the air in or about the Gulf of Mexico; the air thence rising has its place supplied by the next more northern, cooler, and therefore denser and heavier air; that, being in motion, is followed by the next more northern air . . . on a successive current, to which current our coast and inland ridge of mountains give the direction of North-East, as they lie N.E. and S.W.
—Benjamin Franklin

THE BREAKTHROUGH

Scientists had recently begun experimenting with electricity, and Franklin was fascinated by it. He performed many experiments using Leyden jars, which contained and displayed a charge produced by static electricity. When a Leyden jar was charged, a person could bring a finger close to the center rod in the jar and be amazed by the spark that jumped from rod to finger with a sharp cracking sound. Once—using two larger Leyden jars that contained 40 times the charge of an ordinary one—Franklin demonstrated the power of electricity by effectively electrocuting a turkey before cooking it for dinner. (He also nearly electrocuted himself when he became distracted while speaking to a friend and accidentally touched a charged rod. There was a noise as loud as a gunshot, and Franklin twitched violently and blacked out. Luckily, he suffered nothing more than soreness after he regained consciousness.)

Like other scientists before the discovery of the atom, Franklin viewed electricity as a fluid substance. Most scientists believed there were two kinds of electrical fluid, existing in equal amounts in all objects; when an object became electrified, some of one kind of fluid was destroyed, leaving an excess of the other. Franklin, however, argued that there was only one electrical fluid, which was transferred from one object to another but was never destroyed. (Coining a number of terms still used today, Franklin called an object with a normal amount of fluid "neutral," the object that gained fluid "positive," and the

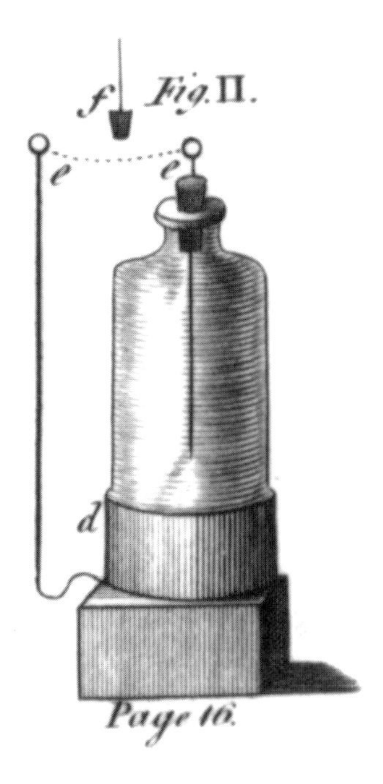

The Leyden jar was invented by Pieter van Musschenbroek of Leyden, the Netherlands, in 1745. It consisted of a glass bottle (which did not conduct electricity) coated with metal. The inside was filled with water or metal shot. A metal rod or wire was inserted into the bottle through an insulating cork. A static charge could be introduced by rubbing the bottle with a fine cloth, usually silk. The charge would remain in the bottle until someone touched the wire. The Leyden jar served as a primitive condenser or capacitor (a device that stores an electric charge).

object that lost fluid "negative." He also was the first to use the word "charge" to describe the presence of electricity.) Through simple experiments in 1747, he had noticed that a charged Leyden jar could be discharged without sparking if he touched it with a pointed metal rod. He correctly concluded that the rod collected, or conducted, the charge from the jar and spread it harmlessly through the air or into the ground.

Scientists now know that lightning is caused by electricity traveling between areas of opposite charge within a cloud, or from a cloud to the ground.

In 1749, Franklin theorized that the electric charge of the Leyden jar was merely a smaller form of lightning. To prove that lightning was a large, rapid release of electrical fluid from the sky, he looked for a way to coax some of this fluid out of storm clouds. His first experimental idea, introduced the following year, involved constructing a "sentry box" large enough to hold a man and placing it atop a tall building. A long, pointed metal rod would extend from the box 20 or 30 feet into the air. When a storm passed over, the rod would draw the electricity from the clouds. The man inside could use a Leyden jar to check for a charge on the rod. Franklin could not perform this experiment himself, since Philadelphia contained no buildings tall enough. (George Richmann, a scientist in Russia, tried the experiment several years later and was killed by a direct hit from a lightning bolt.)

This illustration of the sentry-box experiment was published in a later edition of Franklin's 1751 book Experiments and Observations on Electricity.

A second experiment fixed Benjamin Franklin's image in history. When a hot spell in June 1752 was about to be broken by a gathering thunderstorm, he called to his 21-year-old son, William. William brought a silk kite, atop which he had fastened a

15-inch wire. At the base of the twine of the kite Benjamin had attached a key, and then a silk strip to hold on to. Benjamin hoped that the wire would draw the electric fire from the clouds, that the key would collect the charge, and that the silk—which did not conduct an electric charge—would protect him from being shocked. (Contrary to popular myth, the last thing he wanted was for lightning to actually strike the kite.)

In an open field near their home, William ran through the rain until the kite was airborne. Then,

Benjamin (left) and William Franklin performing the famous kite experiment

taking shelter in a shed while holding the string of the kite, both Franklins waited for something to happen. They were very nearly disappointed. Again and again Benjamin would touch the key—with no effect. Losing hope, he noticed suddenly that the loose threads of the wet twine were standing erect. They had become charged. Sure enough, when he touched the key, there was the familiar crackle and spark. Benjamin picked up a Leyden jar and electrified it using the key. He had successfully proven the nature of lightning.

For the rest of the summer, the Franklins kept their secret while Benjamin wrote a paper on his discovery, which he published for the scientific world that autumn, to great acclaim. (He had, as always, been lucky. The next two people who tried to copy his famous experiment were both electrocuted.) In November 1753, Benjamin Franklin was elected unanimously to the Royal Society and was awarded its highest honor, the prestigious Copely Medal, for his discoveries in electricity. This was followed by many other honors, including honorary degrees from Harvard and Yale. The man who had arrived in a strange town at the age of 17 with less than two years of schooling and little more than the clothes on his back was now known as "Doctor Franklin."

THE RESULT

Franklin's scientific successes continued, due not only to his ability to visualize experiments, but also to his skills as an inventor and gadget maker. He found a practical application for his work in electricity, suggesting that people could avoid lightning damage to their homes and barns by installing pointed metal rods that ran from the roofs of the buildings to the ground. These would conduct the harmful lightning bolts and divert them safely into the earth. At first, many scoffed at the idea—lightning, after all, was still viewed as something beyond anyone's control. But by 1782, there were more than 400 lightning rods on homes in Philadelphia alone. (Always a showman, Franklin set one on his own chimney and connected it to two bells with a tiny brass ball between them. When the rod became electrified, the ball would wobble back and forth, causing the bells to ring and delighting his guests.)

In addition, Franklin improved on the Leyden jar by stringing several of them together, forming an early electric battery. His personal experiences and needs also led to his invention of a more efficient stove he called the "Pennsylvania fireplace," a rocking chair, bifocal eyeglasses, glass chimneys for street lamps that would direct the smoke upward so the glass would remain clear, a rubber catheter, a means of street paving that would supply traction for horses and at the same time a smooth ride for carriage passengers, a musical instrument called the "armonica" (made of wine glasses filled with colored liquid), and

---

Some people were actually offended by Franklin's lightning-rod theory, believing that lightning was the wrath of an angry God and that attempting to avoid it would be interfering with divine will. But Franklin retorted, "Surely the Thunder of Heaven is no more supernatural than the Rain, Hail, or Sunshine of Heaven, against the Inconvenience of which we guard by Roofs and Shades without scruple."

Armonica players held the base of a glass goblet filled with liquid, wet one finger, and lightly rubbed it around the rim until the glass vibrated, producing a tone. Goblets containing different amounts of liquid played different notes. The instrument was so popular that some of the era's greatest composers, including Wolfgang Amadeus Mozart and Ludwig van Beethoven, wrote songs for it. Unfortunately, most goblets at the time were made with lead, and armonica players kept their fingers moist by licking them repeatedly. Several players developed lead poisoning as a result.

> As we enjoy great advantages from the inventions of others, we should be glad of an opportunity to serve others by any invention of ours.
> —Benjamin Franklin

even a variation on his rocking chair that swatted flies as the chair moved (not all of his inventions were instant hits). Franklin's popularity was further increased by his tendency to refuse patent rights for his many inventions, thereby giving his ideas to the people.

Franklin's interest in meteorology never left him. His studies of lightning moved him to work passionately for public safety. He recommended that all ships, as well as homes, be equipped with lightning rods. He also suggested that taking shelter

By 1846, it was common for ships to be protected by lightning rods.

under a tree during a thunderstorm increased the risk of being struck by lightning. In 1763, Franklin was part of a series of discussions on world climate that were hundreds of years ahead of their time, showing that some enlightened people were able to think of issues larger than national boundaries. Franklin theorized that clearing land for farming caused the land to absorb more heat, raising climate temperatures. In a related discussion, he pointed out that climates could be altered and weather patterns changed by volcanic eruptions.

Franklin also suggested changing the rigging on ships to make them safer, offered ideas to improve the diets of sailors, developed "swimming anchors" for use when a ship was in deep water, invented an odometer, advocated iceberg lookouts, and designed water-tight compartments in a ship's hold to prevent sinking.

Weather was most important to sailors, and Franklin loved the sea. During his time as a postmaster, he became interested in the complaint that mail seemed to take two weeks longer to arrive from England than it did to return. In 1769, Franklin spoke to sea captain Timothy Folger about the problem. Folger informed him that colonial captains were familiar with the warm ocean current known as the Gulf Stream, which flowed north from the Gulf of Mexico along the eastern coast of North America and across the North Atlantic to Europe. They used it to their advantage, whereas British captains often sailed against it, slowing their travel. Folger made Franklin a chart of this current, which Franklin refined and published so that all sailors could take advantage of it. Using a thermometer to measure the Gulf Stream more clearly, Franklin found that it was as much as 19 degrees Fahrenheit warmer than the surrounding water. In 1785, after devising a method of measuring water temperatures down to a depth of 100 feet, he established that the Gulf

To measure deep-water temperature, Franklin corked an empty bottle, tied it to a lead-coated rope, and dropped it down into the ocean. As it sank, the increased water pressure would push the cork down and the bottle would fill with water from that depth. When Franklin drew the bottle back up, the change in pressure would return the cork to its place. Measuring the temperature of the water in the bottle, he found that the difference between surface water and deep water in the Gulf Stream could be as much as 12 degrees Fahrenheit.

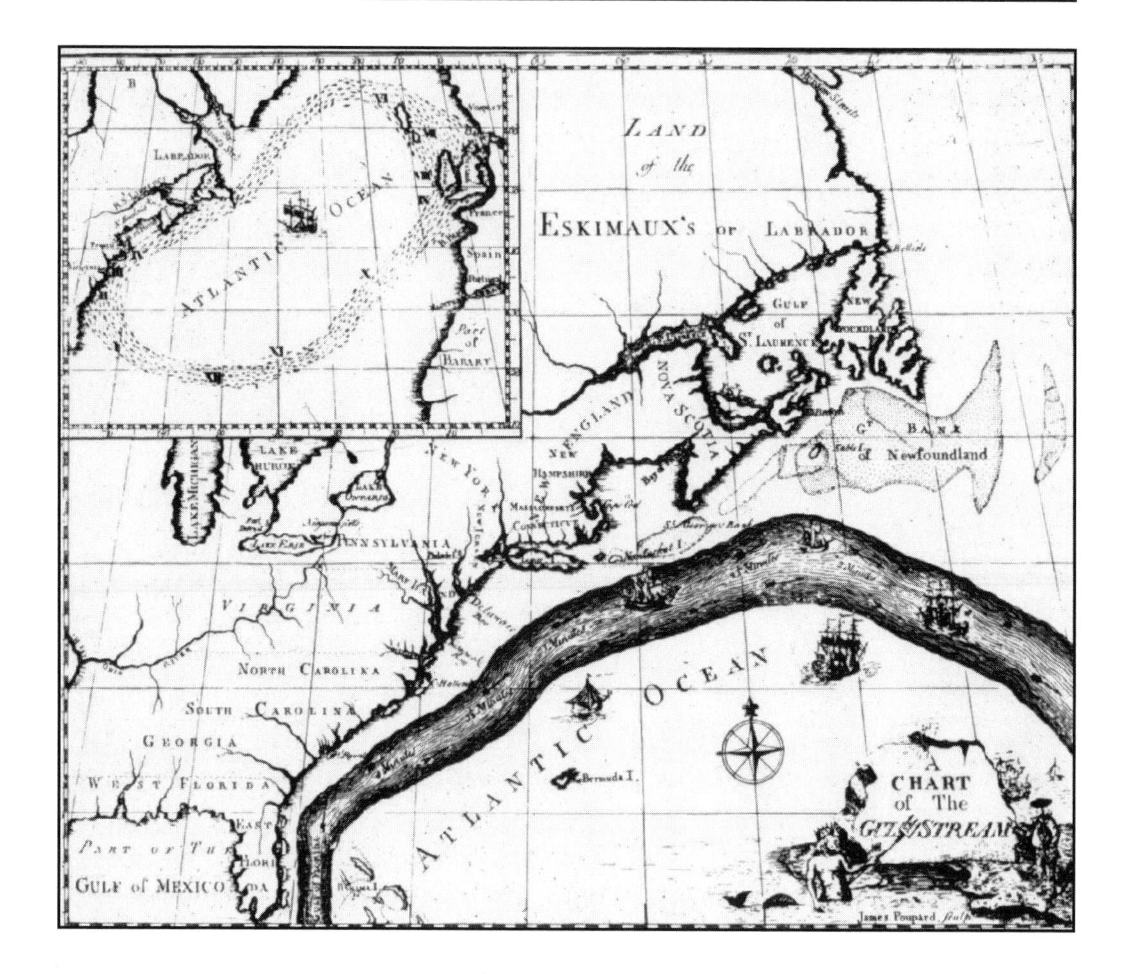

Franklin's map of the Gulf Stream. In the Gulf of Mexico, the current is 50 miles wide and moves at 3 miles per hour (making it one of the strongest currents known). In the North Atlantic, it spreads out to several hundred miles wide and slows to 1 mph.

Stream was like a river of warm water flowing over the colder body of the ocean.

Franklin was a unanimous choice as ambassador to France, and he was instrumental in bringing the French into the Revolutionary War as a U.S. ally. He assisted in drafting the Declaration of Independence, helped negotiate the final peace treaty with England, and was an honored member of the Constitutional Convention. He remained active in politics until his death on April 17, 1790, at the age of 84. Even with all his honors, however, Franklin had thought of himself more as a scientist than a statesman—and as a scientist, his primary goal was to move people from superstition to knowledge. He had proven that lightning was a natural occurrence. It would continue to strike, damage homes, start fires, and kill people, but people would never look at lightning in the same way again.

The signing of the Declaration of Independence on July 4, 1776. In the group of four men standing before the table, Franklin is on the right.

CHAPTER THREE

Luke Howard and Cloud Classification

Almost everyone has, at some time, stretched out on the grass, looked up at the sky, and played the "cloud game." Perhaps a white puffy cloud looks like a rabbit, while long, gray, wispy clouds look like an old man's beard. Throughout history, clouds have been subject to interpretation by myth, religion, art, philosophy, and imagination. Appearing and disappearing, changing shapes, or moving rapidly across the sky, they seem insubstantial and elusive. Yet clouds stand at the center of people's attempts to understand the weather, how it affects their lives, and what it promises for the future.

In early myths, clouds were regarded as the realm of the gods. During the sixth and seventh centuries B.C., however, Greek philosophers began to look at the sky as an area worth studying, thus giving rise to the field of nephology: the study of clouds. Thales of Miletus (625-545 B.C.) knew nothing about evaporation, condensation, or cloud formation, but

Although he had no formal scientific training, Luke Howard (1772-1864) created a system of describing clouds that is still accepted by meteorologists today.

he was able to theorize that water rose and fell between Earth and the heavens, essentially describing the hydrologic cycle. Aristotle (384-322 B.C.) elaborated on this idea, viewing clouds as the perfect example of how everything on Earth is in a constant state of change. He believed that four elements—earth, air, fire, and water—made up the natural world, and their interactions produced weather. For instance, when the Sun's heat (fire) warmed Earth's cold oceans (water), it forced warm, moist air into the sky. When that air reached a certain elevation, clouds formed. For a theory based only on assumptions, it was a good one, and it stood for nearly 2,000 years.

Ideas about the clouds progressed little in the following centuries. French philosopher René Descartes (A.D. 1596-1650) suggested that when clouds became too large to stay up in the sky, they fell as rain, snow, or hail. A popular nineteenth-century view of clouds was the "bubble theory," which stated that when water particles were heated by the sun, they formed bubbles filled with an unusual type of air. This special air was lighter than the normal air in the atmosphere, so the bubbles would rise into the sky and form clouds. Rain or snow would fall when the bubbles burst. The bubble theory gained widespread acceptance, but little was done to study the clouds scientifically to prove or disprove it. Most people believed that clouds were simply too flimsy and fleeting to be analyzed. Ideas about the clouds remained as vague as the clouds themselves, until an amateur meteorologist named Luke Howard invented a new language of the skies.

Luke Howard was born on November 28, 1772, in London, England. His father, Robert Howard, was a successful manufacturer of wrought iron and tinware. Robert's first wife died at a young age, leaving him with three small sons. He soon remarried, however, to Elizabeth Leatham, a woman he had known for years. Luke was their firstborn, followed by three more brothers and a sister.

The Howards were devout Quakers. Robert demanded respect, had little patience for impractical pursuits, and encouraged in his children a hunger for knowledge. When Luke reached school age, his father sent him off to a strict and highly regarded Quaker institution in Burford, near Oxford. The school's headmaster allowed students who were fast learners to set up part of their own curriculum. For Luke, this meant the study of natural history, including field trips into the countryside. He took pleasure in being outdoors in the pastures and hills, under the changing skies.

The Quakers, also known as The Religious Society of Friends, are a Christian group that was founded in around 1650 as a protest against the Church of England. Their teachings were based on the belief there was a divine spirit, or "Inner Light," in every person that would lead him or her to true belief and understanding. Philanthropy and social reform—particularly pacifism, abolitionism, and belief in the equality of all men and women—were integral to Quaker doctrine.

In 1783, when Luke was 10 years old, the summer skies brought panic to the people of northern Europe and Asia. After a temperate spring, the daytime weather became hot and suffocating and the nights bitterly cold. By the middle of the summer, a sickly haze fouled the air, which reeked of sulfur. Leaves fell from the trees in the middle of June. People began having headaches, respiratory problems, nausea, and depression. These events were explained when American meteorologist Benjamin Franklin proposed that the sudden change in climate might be the result of a geological occurrence.

Franklin was correct: four volcanoes in Japan and Iceland had recently erupted, spewing dust, ash, and sulfur into the atmosphere.

Luke gazed eagerly at the strange skies and read accounts written by scientists about the weather they were witnessing throughout Europe. Thus began his lifelong passion of looking up at the heavens. His school lodging was in the back of a building with large windows, and from his desk, Luke began observing the clouds and keeping journals of how they appeared and moved. He puzzled over strange cloud formations, seeking the connections between them and the weather. Not all subjects were as interesting to Luke, however. He had to spend long hours of boring repetition studying Latin grammar. Little did he realize that these exercises would help place his name forever in meteorological history.

When Luke's schooling ended in 1788, he returned to his parents' house and set about working in their garden. He created a small weather station there that included a rain gauge, a thermometer, and a barometer. Twice a day, Luke took readings on wind direction, air pressure, rainfall, and temperature, recording them in a journal. Robert Howard, however, did not want to see his son wile away the hours on such frivolous tasks. He soon sent Luke off to apprentice as a chemist in Cheshire, under the watchful eye of Ollive Sims, a business acquaintance. Cheshire was in northern England, away from London's many distractions; Sims, an orthodox Quaker, was stern and sober. Luke was kept busy from morning to night cleaning the laboratory,

barometer: a device used to measure atmospheric pressure

grinding chemical preparations, and stocking the shelves. There was no time for any leisure activity, and Luke desperately wanted to return to London to pursue a career in science. Being a responsible young man, however, he finished his apprenticeship and returned to London at the age of 21.

Luke soon set up his own pharmacy with money borrowed from his skeptical but supportive father. The shop was small, with cramped living quarters above it, and Luke worked hard. But he was finally free to pursue his scientific curiosities, attending lectures and discussions several nights a week. At one of these, Luke met another young businessman named William Allen. Allen had taken charge of a London pharmaceutical company and was planning to build a larger manufacturing laboratory at Plaistow, not far from the city. He asked Luke to manage the laboratory. The offer couldn't have come at a better time, since Luke was about to marry and needed a larger home. He married Mariabella Eliot on December 7, 1796, and the couple moved to a spacious house in Plaistow soon afterward. They had their first daughter, Mary, in 1797, and six more children would follow in the coming years.

It was Luke Howard's acquaintance with William Allen that brought him closer to his meeting with meteorological destiny. Moving out of the city enabled Howard to begin observing the sky and clouds again. He built a viewing station on the upper floor of his new home, where large windows commanded a view of the wide horizons of the surrounding marshland. Meanwhile, science clubs had

Like Catholics and Jews, Quakers were barred from private schools and universities in England (as well as being forbidden to hold public office). Determined to gain knowledge, they studied independently, founded their own schools, and formed intellectual clubs such as the Askesian Society.

blossomed all over England. Allen and other young Quakers founded their own group in 1796, which they named the Askesian Society. The atmosphere at the club was characterized by a lack of restraint and the belief that no scientific theory was too sacred to be tested. Members read their own papers aloud, performed experiments and demonstrations, and engaged in debate. Their broad range of scientific interests included astronomy, chemical attractions, and explosives. Allen invited Howard to join them.

William Allen (1770-1843) would later gain his own scientific fame as a speaker and a member of the Royal Society.

THE BREAKTHROUGH

On a cold night in December 1802, Luke Howard nervously rose before the Askesian Society and read a paper entitled "On the Modifications of Clouds." Based on his sketches and journals from many years of observation, the essay presented his nomenclature, or set of names, for clouds.

There had been earlier scientific attempts to name the clouds. In 1665, for example, Robert Hooke (1635-1703) devised a language to describe clouds as "faces of the sky." His system, however, was loose and unsystematic, using terms such as "hairy" (small, thin, high clouds), "waved" (large, low clouds), or "thick" (a sky whitened by many vapors). Hooke eventually lost interest, pursued other ideas, and became disengaged from meteorology.

Another effort was made by the Societas Meteorologica Palatina at Mannheim, Germany. In the 1700s, the society established 50 or more weather stations extending from North America across Europe to Siberia. The people working at these stations needed a standard vocabulary to record and communicate their weather observations. What resulted was a cloud-naming scheme that used combinations of terms (one set describing color and another describing shape) to reflect the fact that clouds can change and move into one another. Unfortunately, the French Revolutionary Army disrupted the society's efforts when it destroyed the city of Mannheim in 1795.

"It is not in the least amiss for those who are involved in meteorological research to give some attention to the form of clouds," Lamarck wrote, "for, besides the individual and accidental forms of each cloud, it is clear that clouds have certain general forms which are not all dependent on chance but on a state of affairs which it would be useful to recognize and determine."

Shortly before Howard presented his paper, another man, Jean-Baptiste Lamarck (1744-1829), developed his own cloud-naming system. Lamarck correctly suggested that clouds did not form by chance and that they could be classified according to altitude, which he designated as high, middle, or low. But his terminology, like those before it, was loosely defined. It simply described different clouds based on their shape, color, and texture, rather than dividing them into groups by how they formed and behaved. Furthermore, Lamarck's system was expressed in French, which was not the language of the scientific world. These flaws—along with the fact that France's leader, Napoleon Bonaparte, discouraged his efforts—put an end to Lamarck's work in meteorology.

Luke Howard shared Lamarck's insight that although clouds may have thousands of individual shapes, they have only a few basic forms. He went even further, however, by realizing that both the shapes and the forms of clouds are caused by specific physical conditions (such as air temperature, humidity, and pressure) that affect the water present in the atmosphere. In his paper, Howard disputed the bubble theory of cloud formation, correctly suggesting that clouds form when warm air rises and cools to the dewpoint—the temperature at which water vapor condenses into liquid. The vapor condenses around natural particles in the air (such as dust, pollen, or sea salt) into solid droplets of water, billions of which form a cloud. Clouds in colder air, he noted, might also contain ice crystals, giving them a wispy look. In contrast, since warm air holds more water vapor, hot and sunny days would tend to produce larger, puffier clouds.

Based on these observations, Howard was able to categorize clouds in a way that provided clearer distinctions among the different types. He divided them not only by their appearance and altitude, but also by how they were formed. Howard's method was inspired by the work of Swedish botanist Carl von Linné (1707-1778), known as Linnaeus, who had earned fame with his system of classification for all plants and animals. Linnaeus grouped organisms broadly into large groups, then more specifically into smaller groups. (For example, the genus "Canis" could be further defined by the species "familiaris" to identify the domestic dog, while "Canis lupus" identified the wolf.) All the groups were named in Latin, the language of science and scholarship. It made perfect sense to Howard that a similar nomenclature could apply to the clouds.

Linnaeus's name system for plants and animals has remained the basis for modern taxonomy, the science of classification.

Howard presented three basic groups into which every individual cloud formation could be categorized. Each group had a Latin name that reflected its appearance. (With this innovation, Howard's hours of studying Latin grammar finally paid off.) The first was cumulus (Latin for "heap"), a low-level, fluffy cloud with what he called "convex or conical heaps, increasing upward from a horizontal base." Second was cirrus (Latin for "hair" or "fiber"), a high-level, wispy cloud with "parallel, flexuous, or diverging fibers, extensible in any or in all directions." Third was stratus (Latin for "layer"), a

One of Luke Howard's watercolor cloud studies from about 1810, showing puffy cumulus clouds mixed with a few flat stratus clouds

low-level, flat cloud that looked like "a widely extended, continuous, horizontal sheet." Offering his sketches as illustrations, Howard boldly stated that each of the basic cloud forms was "as distinguishable from each other as a tree from a hill, or the latter from a lake."

To address the fact that clouds often unite or disperse, Howard named four other cloud types that were modifications of the major groups. Cirrocumulus was a high-level cloud that formed in heaps or lumps, while cirrostratus was a high-level cloud that formed in horizontal or slightly inclined layers. Cumulostratus was a stratus blended with a cumulus, and nimbus was a rain cloud. With these categories, Howard's system emphasized that even though clouds changed, they always changed in recognizable stages.

Howard also stressed that clouds could be read as visible signs of large atmospheric processes. His essay included comments on how clouds could be

This sketch by Howard shows wispy cirrus clouds (top), along with cirrostratus (right) and cirrocumulus clouds (bottom).

used to predict the weather. He observed, for example, that "continued wet weather is attended with horizontal sheets of [cirrus clouds] which subside quickly and pass to the cirro-stratus." Of cumulus, he said, "The formation of large cumuli to leeward in a strong wind, indicates the approach of a calm with rain." Howard even developed a system of shorthand symbols for rapid notation of cloud observations by meteorologists (which remains in use, in a modified form, today).

It was clear to everyone attending Howard's lecture that no one else had ever accomplished the feat of naming the clouds with such precision and clarity. One particularly impressed audience member was Alexander Tilloch, a publisher and magazine owner. Tilloch asked Howard to contribute his ideas on clouds to the *Philosophical Magazine*, the best-known science journal in Europe. Howard expanded and revised his paper and clarified his drawings, and "On the Modifications of Clouds" was published in the magazine in three parts in July, September, and October 1803. The revolutionary theory was now available for the scientific world to read and discuss.

There is much more to be done in the field, as I have sought to intimate already, but these, at least, are the clouds as I know them. Or, perhaps I should say, these are the clouds as I have so far understood them.
—Luke Howard, from "On the Modifications of Clouds"

THE RESULT

Over the next decade, Howard's essay and excerpts from it were reprinted in many journals. He wrote a monthly weather column in which he described the weather using his new terminology and encouraged other meteorologists to do so. He also maintained his habit of keeping daily weather observations, which themselves proved to be profitable. Howard compiled his data into a 700-page book entitled *The Climate of London*, published in two parts in 1818 and 1820. This book became a classic in the study of urban meteorology. It was the first to demonstrate that conditions in cities, such as the emissions from burning coal, could alter the local weather. Howard called the phenomenon "city fog," which is today known as smog (a mixture of "fog" and "smoke").

Howard's cloud theories soon found a wide variety of uses. Some were practical: William Scoresby Jr., the young captain of a whaling vessel sailing to Greenland in the summer of 1810, used the new nomenclature to describe weather conditions in the ship's log. Other uses were artistic: knowledge of the clouds aided Romantic landscape painters such as John Constable, who became famous for his depictions of the English skies. Constable was so inspired by Howard's theories that he painted more than 100 cloud studies in the summers of 1821 and 1822 alone.

As word of Howard's system spread, it captured the attention of Johann Wolfgang von Goethe, a prominent intellectual figure who exercised great

A series of speeches that Howard made in 1817 was published in 1837 as *Seven Lectures in Meteorology,* thought to be the first-ever meteorological textbook.

The Climate of London also pointed out that temperatures are warmer at night in urban centers than in the surrounding countryside. This phenomenon, well accepted today, is now called the "urban heat island."

While he was gaining recognition, Howard was ending a working relationship with William Allen. In 1805, Howard moved the Plaistow pharmaceutical laboratory to a larger building in nearby Stratford. Two years later, Allen agreed to start a smaller company in London while Howard's factory became independent as Luke Howard & Company of Stratford. The split was amicable, however, and Howard and Allen continued their strong friendship.

John Constable was not the only landscape painter inspired by Howard's work. This 1849 watercolor by E. Kenyon incorporated some of Howard's cloud studies.

influence over Europe's literary culture. Goethe used the Howard system in his own weather journals, and his word settled the controversy over the translation of Howard's nomenclature into German. Even though Latin was the global language of classification, most scientific writing was still translated into local languages. Goethe, however, believed that Latin names should be accepted worldwide without translation, and so Howard's cloud names remained in Latin throughout Europe. In 1821, Goethe published poems on each of the cloud types under the general title "In Honor of Howard." He also wrote

Howard a letter requesting information about his life and his work. Howard was so taken aback that someone of Goethe's stature should want to know about him that at first he thought the letter was a practical joke, but he answered with a brief memoir and a copy of *The Climate of London*.

Famous and financially secure, Howard was elected to the Royal Society in 1821, and in 1823 he became a founding member of the Meteorological Society. But although he valued his contribution to science, the fame that came with it did not set well with his Quaker convictions. He began to withdraw from London life after buying property in Yorkshire, where he volunteered as a teacher at a Quaker school, campaigned against slavery, and helped fund relief for victims of war. Gradually, his attendance at the meetings of the Meteorological Society ceased. Howard also left his daily weather recordings to members of his family. After his wife died in 1852, he lived with his eldest son, Robert, in Tottenham. On March 21, 1864, Luke Howard died at the age of 91. He was buried next to his parents, his wife, and two of his half-brothers, under the English country sky that he loved so much.

Meteorologists outside Europe were slower to adopt Howard's nomenclature, but his cloud names eventually caught on internationally. New terms were added, bringing the number of recognized basic forms to 10. Today, all classifications are based on Luke Howard's original system, giving scientists around the world a common language in which to discuss the clouds.

But Howard gives us with his clearer mind
The gain of lessons new to all mankind;
That which no hand can reach, no hand can clasp,
He first has gain'd, first held with mental grasp.
Defin'd the doubtful, fix'd its limit-line,
And named it fitly.—Be the honor thine!
As clouds ascend, are folded, scatter, fall,
Let the world think of thee who taught it all.
—Johann Wolfgang von Goethe

Those who lived with him will not soon forget his interest in the appearance of the sky. Whether at morning, noon, or night, he would go out to look around on the heavens, and notice the changes going on. . . . Long after he ceased, from failing memory, to name the "cirrus," or "cumulus," he would derive a mental feast from the gaze, and seem to recognize old friends in their outlines.
—Robert Howard's funeral speech for his father

CHAPTER FOUR

Vilhelm Bjerknes and Fronts and Cyclones

A storm moved quickly eastward over the North Atlantic Ocean on October 22, 1921. The forecaster for the Danish Meteorological Institute in Copenhagen didn't think much of it, however. Since the storm seemed to be weakening, he gave the order to remove the warning signals for the north coast of Denmark. The following day, Copenhagen was calm and everything seemed normal. No news of weather problems arrived at the Institute.

Unfortunately, the reason for the lack of word from the north coast was that communication lines were down as a result of hurricane-force winds. Fishing boats and ships in the area were destroyed and many people were killed. In the face of public outrage, the head of the Meteorological Institute claimed that the storm could never have been predicted. But on the neighboring coasts of Norway and Sweden, fishermen had been warned to stay on land, boats were lashed securely, and coastal towns

A reluctant meteorologist at first, Vilhelm Bjerknes (1862-1941) went on to revolutionize the science with his forecasting methods.

rode out the storm safely. Meteorologists there had accurately predicted the storm because they had all been trained in the forecasting methods developed by Vilhelm Bjerknes.

The scientific accomplishments of Vilhelm Bjerknes actually began with his father, Carl, and continued with his son Jacob. It is the story of three generations of scientists—and at the center of this dynasty is Vilhelm Bjerknes, often called "the father of modern meteorology." Vilhelm was born on March 14, 1862, to Carl and Aletta Bjerknes in Christiania, Norway. (Christiania was spelled Kristiania after 1877 and was renamed Oslo in 1925.) Carl Anton Bjerknes was a scientist who suffered from a flawed reputation. He had formulated a theory of hydrodynamics, the branch of physics that studies the motion of fluids. He was not very good, however, at arranging experiments that would support his ideas or show how they might apply to real situations. Carl also tended to be a loner—even when he could have used the support of colleagues.

From an early age, Vilhelm felt a responsibility to help his father and continue his work. He entered the University of Kristiania to study science in 1880, earned a master's degree in 1888, and collaborated with his father throughout his schooling. After they had been working together for several years, however, Vilhelm began to worry that continuing the partnership could harm his own scientific reputation. He realized that the scholarly community would not accept a scientist who operated in isolation. He still wanted to help his father organize and publish his work, but he needed to find his own career path.

Vilhelm went to Paris to study abroad on a fellowship in 1889. While he was there, he followed the emerging field of electrodynamics, the branch of physics that deals with electric charges and currents. He became acquainted with the work of Heinrich Hertz, who had recently shown the existence of electromagnetic waves. The behavior of these waves was similar to the behavior of fluids, and Vilhelm hoped that the great attention given to Hertz's work

electromagnetic waves: waves made up of an electric field and a magnetic field that have the same frequency and travel at the speed of light (kinds of electromagnetic waves include radio waves, X-rays, and ultraviolet rays)

Heinrich Rudolf Hertz (1857-1894) determined that electromagnetic waves followed the same physical properties as light waves did: they could travel through the air and be transmitted, reflected, and received. In 1895, just after Hertz's death, Guglielmo Marconi (1874-1937) successfully transmitted an electromagnetic signal over a distance, inventing the wireless telegraph. This technology aided the development of meteorology by allowing observations and forecasts to be communicated rapidly around the world.

Bjerknes remained a friend of Hertz and his family. In fact, more than 40 years later, Bjerknes helped Hertz's widow and daughter escape from Nazi Germany.

might also bring recognition to his father's hydrodynamic theories. In 1889, however, concern arose over the validity of Hertz's research. Vilhelm was shocked at the challenge to this famous scientist, whose failing eyesight prevented him from developing new experiments to defend his theories.

Vilhelm Bjerknes wrote to Hertz to ask if he could study in his laboratory at Bonn University in Germany, and Hertz agreed. When Bjerknes arrived in Bonn, he was startled to find the lab in disrepair. It turned out he was Hertz's only assistant. For the next two years, he worked to help Hertz clear his name. Grateful, Hertz encouraged Bjerknes to stay on in Germany, where there were more scientific opportunities than in Norway. At his father's urging, however, Bjerknes returned home.

Although the time spent with Hertz had kept Bjerknes from advancing his own career, he was able to earn his Ph.D. in engineering in 1892, basing his thesis on his work in Bonn. This was followed by appointments as lecturer at the Högskola (engineering school) in Stockholm, Sweden, in 1893, and then as a professor of applied mechanics and mathematical physics at the University of Stockholm in 1895. Having found a comfortable position, he followed his plan of completing and publishing his father's research. The stress of trying to link Carl's work in hydrodynamics with his own interest in electrodynamics nearly proved too much for him, however. Colleagues were afraid that he was nearing a nervous breakdown. Finally, in 1902, he finished the second of two volumes based on his father's life work.

They were not given the attention either Vilhelm or Carl Bjerknes thought they deserved, but Vilhelm's debt to his father had been paid.

The energy Vilhelm directed to his father's work had kept him from following shifts in the scientific community. Vilhelm learned science during a time when most physicists believed the world could be explained mechanically, as if it were a complex machine. His work so far had been based on that view. The discovery of atomic particles, however, caused many to believe that instead the world could best be explained through the workings of atoms and molecules. Vilhelm clung to his mechanical focus and tried to defend it, but few other scientists agreed with him. Although he had seemed a promising young scientist in 1900, by 1905 Vilhelm Bjerknes was out of the mainstream and heading toward obscurity.

In fact, the same could be said for the field of meteorology: it was not highly respected. Attempts at predicting the weather were little more than educated guesses, and the meteorological community often accepted work that was sloppy, lazy, or poorly planned. Two pieces needed to fall into place to help Vilhelm Bjerknes turn around not only his own career, but also the entire science of meteorology—and they were looming on the horizon like gathering storms. Both events would change the world forever; one would be considered the most exciting adventure of the new century, while the other would be one of its worst disasters.

THE BREAKTHROUGH

At the beginning of the twentieth century, the world was fascinated by new technology, and no invention promised greater excitement than the ability to fly. Hot-air balloons had been in use since the late 1700s, and dirigibles (balloons that could be directed or steered) were nearly 50 years old. Inventors and daredevils kept seeking better ways to fly, however, and their enthusiasm was contagious. By 1903, when the Wright brothers took to the air in the world's first airplane, the public was already imagining the possibilities of flight. All this interest and progress in aviation made the atmosphere an increasingly popular focus of scientific investigation. Aircraft flew at high altitudes, and pilots needed to be able to predict air movements so they could stay on course and avoid bad weather. At the same time, aircraft could aid meteorology by gathering new data about temperature, pressure, and other conditions in the upper atmosphere. (Traditionally, weather predictions had been based solely on observations of conditions on the ground.)

Bjerknes did not see an immediate connection between manned flight and his work. He had been focusing on mathematical explanations of how vortices, or swirling movements, formed in liquids. But in an 1898 lecture on hydrodynamics, he remarked that his theories might be applied to movements in the atmosphere. His background in mechanics led him to view the atmosphere as a massive engine that converted the sun's heat energy into mechanical energy that moved the air. When the sun heated the atmosphere, the warm air rose over the colder air, creating vortices. Bjerknes realized that his knowledge of how vortices formed and circulated could help scientists predict air movements—and, therefore, predict the weather.

One of Bjerknes's colleagues in Stockholm was Nils Ekholm, a meteorologist interested in hot-air ballooning and polar exploration. Ekholm was among a number of Stockholm scientists who wanted to apply the methods and laws of chemistry and physics to studying the earth, seas, and atmosphere. He had already demonstrated that the air sometimes behaved like a liquid, and he was intrigued by the way Bjerknes's theories could apply to meteorology. Bjerknes had suggested in his lecture that observations of the upper air gathered by balloons or kites might allow him to use his hydrodynamic equations to study atmospheric movements. So Ekholm worked to establish a committee at the Stockholm Physics Society to collect this information, inviting Bjerknes to join the effort. Bjerknes struggled with the feeling of, as he called it, "getting sucked into the meteorological vortex," but he accepted the opportunity. As a theoretical physicist, he had been largely ignored. Focusing on practical applications such as meteorology, however, offered more funding, acceptance, and even prestige.

While his career was beginning to take form, however, his personal life was falling apart. In 1903, Carl Bjerknes died suddenly. Then over the next two years, Vilhelm's oldest son, Karl, required a

> I could no longer withdraw from the answer to that question: What do I really want [to do]? I found the answer could be only one: I want to solve the problem of predicting the future states of the atmosphere and ocean. I had previously closed my eyes to the fact that this actually was my goal, I must confess, partially for fear of the problem's enormity and of wanting too much.
> —Vilhelm Bjerknes, from a letter to a colleague, 1904

series of operations; his younger sons suffered from stress-related disorders; his wife, Honoria, had a miscarriage; and her father died. Vilhelm had little money and few friends, and he dealt with frequent insomnia and depression. Perhaps out of a need to commit himself to something permanent, he decided to devote his professional life to meteorology. By doing so, he could defend his mechanical view of physics while influencing both science and society.

Weather forecasts had previously been based on patterns experienced over time. If a storm arrived in city A one day and in city B the next, forecasters assumed that the weather for city B would probably always be the same as that of city A, only a day later. Cloud patterns indicated when rain might be in the forecast, and temperature and air pressure changes gave a good indication of what was about to happen—but none of these methods were of much use for a forecast that extended beyond 24 hours.

Bjerknes wanted to take a new approach to forecasting by using physics. He determined that the movements of the atmosphere were defined by five variables: temperature, humidity, pressure, wind speed, and air density. Understanding the relationship among these variables would improve forecasts. This effort required two pieces of information. One was the state of the atmosphere at a given time, which Bjerknes compared to a doctor's diagnosis of a patient. The second was a knowledge of the physical laws that made one atmospheric state develop into another—similar to a doctor's knowledge of the symptoms and courses of various illnesses. With this information, forecasters could create charts that would give a "prognosis" of what would happen next, just as a doctor predicted the fate of a sick patient.

In 1905, Bjerknes visited the United States, where he presented lectures explaining how he hoped to apply mathematical precision to weather prediction. The Carnegie Foundation found his vision intriguing, and in 1906 it awarded him a grant to hire assistants and establish his research. (He would receive this grant for the next 36 years.) In 1907, Bjerknes accepted a position as chair of applied mechanics and mathematical physics at the University of Kristiania.

Meteorology becomes exact to the extent that it develops into a physics of the atmosphere.
—Vilhelm Bjerknes

Bjerknes's work was given another boost in 1912, when he was offered the chance to establish and direct the Leipzig Geophysical Institute in Germany. Germany had emerged as the first nation to express interest in air travel (in fact, Leipzig would soon become a major center for zeppelin dirigibles). An air industry required accurate weather forecasting to ensure the safety of its airships and passengers. This was the sort of scientific work Bjerknes longed for—research that met a practical need. In addition, as an educational and research center, the Geophysical Institute would provide him with talented students who could join his effort and spread his methods to other countries. Perhaps because he remembered his father's struggle with isolation, Bjerknes wanted to gather a large circle of collaborators for his research. One of these would be his son Jacob, who joined the group in 1916 at the age of 19.

In 1900, Ferdinand Zeppelin (1838-1917) built his first airship, a rigid dirigible with an internal frame. Zeppelins, as they became known, were used in World War I for reconnaissance and to bomb England. After the war, they were used for passenger travel, but crashes and the development of faster, safer airplanes brought the zeppelin travel industry to an end by 1940.

When he began collaborating with his father, Jacob Aall Bonnevie Bjerknes (1897-1975) launched a long and distinguished career in meteorology. He is shown here analyzing a weather map in Bergen, Norway, in about 1922.

The collaborative effort at Leipzig quickly produced results. Bjerknes's team published a series of detailed charts that analyzed data from several layers of the atmosphere. The group investigated the effect of mountains on air motions near Earth's surface, finding that airplanes could experience dangerous turbulence if they flew too low over the peaks. It also analyzed the North Atlantic trade winds, mid-level air currents that could be navigated to speed flights between Europe and North America. It even recommended a standard unit of measurement for air pressure, the millibar (still used today), and employed it to map low- and high-pressure fields in the atmosphere.

Then the last piece of the puzzle that would guarantee Vilhelm Bjerknes's success fell into place. The team was hard at work in Leipzig on June 28, 1914, when a Serbian nationalist shot and killed Archduke Francis Ferdinand of Austria-Hungary. Within a few months, France, Great Britain, and Russia had joined Serbia in a war with Austria-Hungary, the Ottoman Empire, and Germany. World War I was devastating, but it helped move the science of meteorology forward. Accurate knowledge of the weather was essential to both sides in deciding when to launch poison-gas attacks, how to aim long-range artillery, and whether zeppelins could safely fly reconnaissance or planes could bomb enemy positions.

Food and fuel shortages eventually made work in wartime Germany difficult. In 1917, Bjerknes moved back to Norway (which remained neutral during the war) to head a new meteorological institute in Bergen. Although he had lost assistants and students on the battlefield, new colleagues joined the effort to improve weather forecasts. Forecasting in Norway had previously depended upon weather data sent by telegram from other countries, but such information was kept secret during the war. The Norwegian navy had set up observation posts to keep watch for German submarines, and Bjerknes realized that these could also serve as weather observation stations. In fact, they were already manned by sailors and fishermen—most of whom were experienced

The more aerial navigation develops to play a role for mankind—in war or peace—the greater will be the demand to know at any time the state and motion of the atmosphere so that aerial voyage can be planned on the basis of this knowledge. And on the other hand meteorology for its development is completely dependent on aeronautics in this word's broadest meaning. It alone can provide the observations that will allow us to study completely the atmosphere's laws.
—Vilhelm Bjerknes, 1909

Bergen was an exciting spot for weather observing. Located in the path of 80 percent of the storm systems from the North Atlantic, it is known as "the wettest city in Europe." Rain falls two out of every three days in Bergen, for a total of nearly 82 inches per year.

Vilhelm Bjerknes started his forecasting effort, the West Norway Weather Bureau, in the attic of his home in Bergen. In this picture, probably taken around 1919, Tor Bergeron (center) and Jacob Bjerknes (right) are hard at work making weather maps.

weather watchers. By the time the war ended in 1918, Bjerknes had set up 60 new weather stations in West Norway. The information they gathered was used to make forecasts that aided not only national defense, but also agricultural production during wartime food shortages.

In the course of observing cyclones (low-pressure systems of storms) along the west coast of Norway, Bjerknes's Bergen group noted that not every storm was a separate entity. Sometimes additional cyclones developed out of and followed the primary ones. Based on the patterns of storms in the area, the group suggested that there was a line of "discontinuity" in the atmosphere of the Northern Hemisphere where cold air moving down from the North Pole met warm air moving up from the Equator. As these opposing air masses met, storms formed "like pearls along a string," as Bjerknes put it.

In naming this boundary, Bjerknes called upon images impressed into everyone's minds during World War I. Both sides had dug lines of deep trenches to shelter soldiers from devastating new weapons such as machine guns. The places along the opposing lines where troops met and fought were called fronts, and they slowly moved back and forth

A collision between warm and cold air masses is visible here as a line in the clouds.

air mass: a large body of air within which the temperature, pressure, and humidity are constant at all altitudes

as first one side and then the other gained the advantage. Viewing the struggles of air masses as similar to the struggles of great armies, Bjerknes called the line between polar and tropical air masses the polar front. He also created names for the various behaviors of air masses at different points along the polar front. If warm air was moving in on cold air, it was called a warm front. As the warm air gradually rose above the cold, clouds would thicken and rain was likely. If cold air was moving in on warm air, it was a cold front. The warm air would be forced to rise rapidly, creating powerful thunderstorms. And if the air masses butted against each other without moving, it was called a stationary front. Weather along it tended to be clear or partly cloudy.

The Bergen group determined that cyclones formed when a wavelike "kink" of low pressure developed in the polar front, with a cold front to the west of it pushing south, and a warm front to the east pushing north. Forced in opposite directions by the two fronts, the air around the low-pressure center began to rotate counterclockwise. Wind speeds increased as air converged toward the center, and the rising of warm air and sinking of cold air generated precipitation. Jacob Bjerknes formulated a basic cyclone model in the autumn of 1918, and the Bergen group then worked to refine his theories.

Using data gathered from different levels of the atmosphere, the group developed a three-dimensional cyclone model that allowed meteorologists to more clearly predict the weather based on atmospheric conditions, such as temperature and pressure, along the polar front. Although the model was initially opposed by scientists unused to thinking of air movements in three dimensions, Vilhelm Bjerknes's colleagues and students continued to spread his theories internationally. By the end of World War II, most of the world's meteorologists were using the methods of the Bergen group, and the polar front theory was accepted as a fundamental concept of meteorology.

A diagram of horizontal and vertical views of a cyclone, as published in a paper by Jacob Bjerknes in 1922.

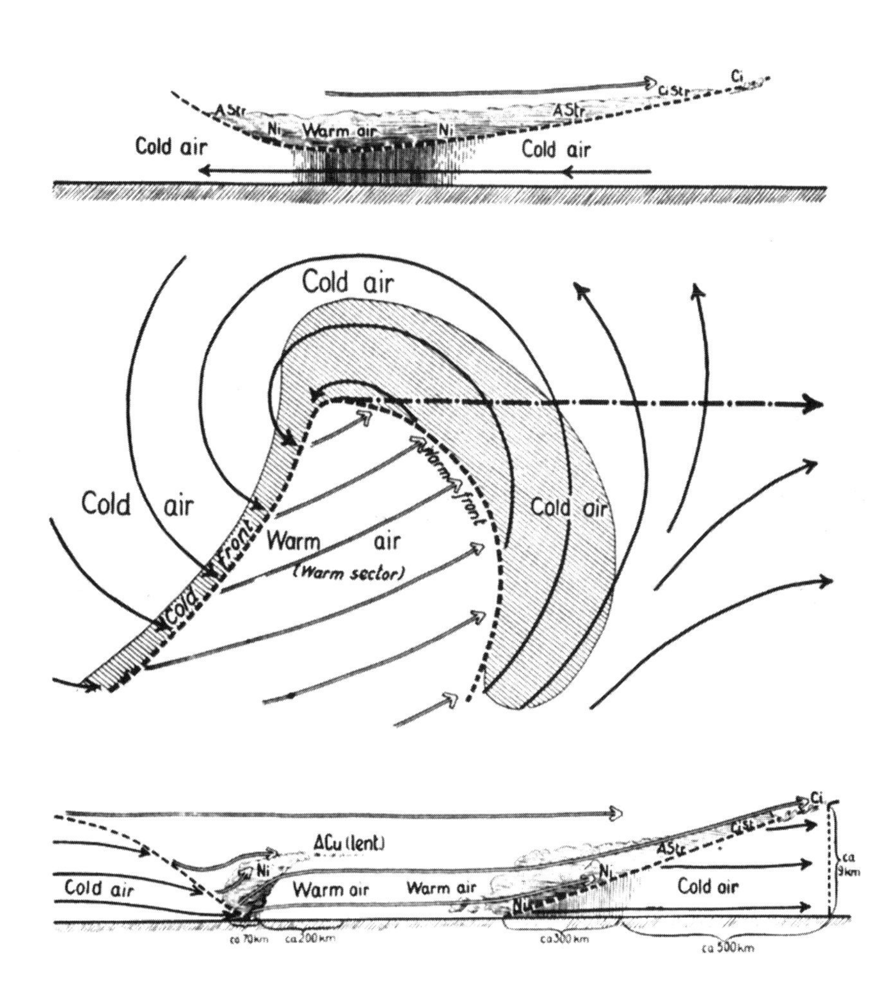

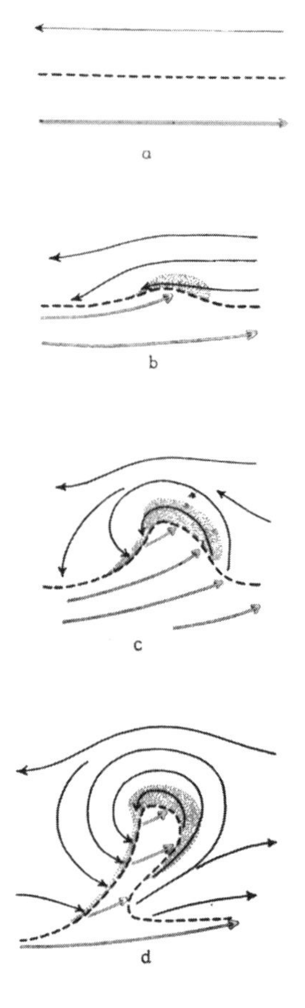

Jacob Bjerknes's model of the formation of a cyclone. Two air masses meet (a), a pocket of low pressure forms (b), and the air begins to rotate around it (c and d).

THE RESULT

Realizing that his scientific projects were subject to "market forces," Bjerknes had learned the skill of altering his research to suit the needs of the time. After World War I ended in 1918, he focused the Bergen group on weather forecasting for farmers and the newly developing commercial aviation industry. A period of economic depression in Norway also prompted the Bergen group to work on forecasts designed for fishermen, since fishing was a vital part of the nation's economy. This resulted in greater knowledge about the life cycle of cyclones.

Bjerknes's team found that as cyclones moved eastward, the faster-moving cold front would overtake the warm front, forming what was known as an occluded front. At this point, the storm would be at its most intense. But since cold air now completely surrounded the low-pressure center, the storm would lose the energy provided by rising warm air. The cyclone would die out and gradually dissipate (although sometimes a new wave would form on the westward end of the cold front, starting the process again). This model, though not absolutely predictable, allowed meteorologists to forecast weather patterns further ahead than had previously been possible. By 1925, the Bergen group had developed a way to classify air masses based on their life cycles, making it possible to predict their behavior with even greater accuracy.

Vilhelm Bjerknes became chair of applied mechanics and mathematical physics at the

University of Oslo (formerly Kristiania) in 1926 and continued there until his retirement in 1932. He died in Oslo on April 9, 1951, having given credibility to the entire science of meteorology. The rise of air travel, key collaborations with his father and his son, and a bloody war had moved him to the height of the scientific establishment. He had established the pattern for the modern research team and learned how to adapt science to public issues. He also understood that the future of meteorology as a science would depend on the next generation, and he worked to educate others in his methods.

Jacob Bjerknes moved to the United States in 1939 and dedicated the rest of his career to teaching his father's theories to American meteorologists. Another of Vilhelm Bjerknes's students would also move to the U.S. and work to establish that country as a leader in meteorological research. Carl-Gustaf Rossby would make his mark on the world by studying the global air currents that propelled Bjerknes's fronts and cyclones. And he would do it by following one of Bjerknes's central plans: the establishment of an influential group of scientists and students.

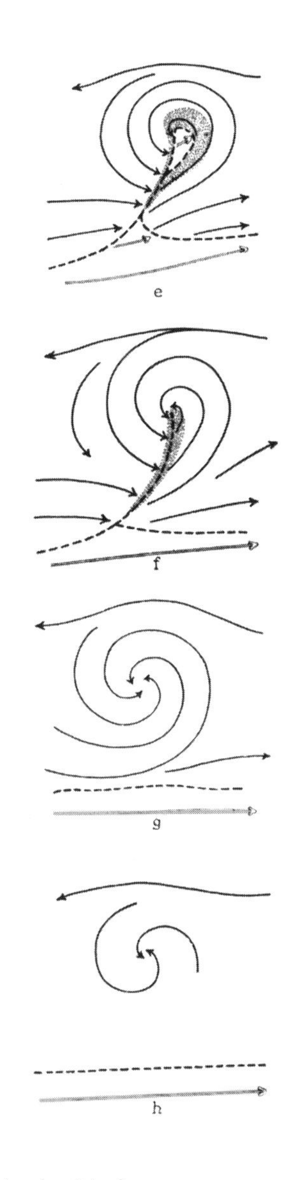

As the Bjerknes team discovered, the cold air in a cyclone gradually overtakes the warm air and "squeezes" it out (e and f), causing the storm to dissipate (g and h).

CHAPTER FIVE

Carl-Gustaf Rossby and Rossby Waves

During World War I, Vilhelm Bjerknes's team of meteorologists was making revolutionary discoveries about the weather. Carl-Gustaf Rossby, then a student at the University of Stockholm, was not interested, however. He was 19 years old, restless, and in search of an exciting profession. With all the world in turmoil, adventure was everywhere, and the possibilities were limitless. Studying maps, recording data, and working out mathematical weather predictions didn't seem to be the stuff of glory. But in time, Rossby would find both glory and excitement in meteorology. Like Bjerknes, he would contribute groundbreaking new ideas to weather prediction—and they would come in a time of war.

Born on December 28, 1898, in Stockholm, Sweden, Carl-Gustaf Arvid Rossby was the son of a construction engineer, Arvid Rossby, and his wife, Alma. Even as a child, Carl revealed the energetic personality that would mark his life and his career.

In studying the movements of air masses, Carl-Gustaf Rossby (1898-1957) aided the airline industry and the military, as well as the science of meteorology.

He went through school easily, without showing any special interest in science, and entered the University of Stockholm in 1917. For a brief time, he thought he would like to be an astronomer, but this attraction came mainly from novels that portrayed astronomers as romantic, bearded loners sitting on mountaintops and gazing at the sky. He soon realized that astronomy would not really be so exciting.

Rossby studied physics for a year and then transferred to the Geophysical Institute at Bergen, Norway. It was there that Bjerknes and his colleagues were performing their research on weather fronts and cyclones. Although Rossby had only been a halfhearted student of physics, he became hooked on meteorology. Soon he was living in the Bjerknes home and listening to every word from his newfound mentor. After absorbing all of Bjerknes's teachings about the movements of air masses, Rossby studied for a year at the University of Leipzig and then returned to Stockholm to earn his graduate degree in theoretical physics. He began working for the Swedish weather service, but found few career prospects there.

Rossby later remembered Bjerknes as "a man with a bushel of hair, a remote interest in his students and a frugal way with his family."

The U.S. Weather Bureau was organized in 1870 as the U.S. Army Signal Service, then became the Army Signal Corps in 1880. In 1891, it was reorganized as the Weather Bureau under the Department of Agriculture. Since 1970, it has been part of the National Oceanic and Atmospheric Administration (NOAA). The bureau was renamed the National Weather Service (NWS) in 1973.

Luckily, Rossby's status as one of Bjerknes's most prized and beloved pupils soon earned him a one-year fellowship to bring the Bergen group's new forecasting methods to the United States. In 1926, he joined the staff of the U.S. Weather Bureau in Washington, D.C. At more than 50 years old, the Weather Bureau had become a very conservative organization, and many viewed it as hopelessly outdated. It was hoped that "new blood" from Bjerknes's

The Washington, D.C., Forecast Office of the U.S. Weather Bureau in 1926. Observers at weather stations around the country would send their readings to the Washington office, where meteorologists plotted charts and drew maps showing changes in temperature, pressure, wind speed and direction, and humidity.

famous team would revitalize the bureau, but that wasn't what happened. American meteorologists didn't much care for new ideas, and they certainly didn't much care for the energetic new guy from Sweden. It was not long before Rossby found himself in trouble. In 1927, pilot Charles Lindbergh earned international fame for his solo crossing of the Atlantic Ocean, then began preparing for a solo flight to Mexico. Rossby used the methods he had learned in Norway to prepare a weather forecast for the famous aviator. It was an accurate forecast, but it was unauthorized. Many at the bureau saw it as shameless grandstanding, an attempt to cash in on Lindbergh's celebrity. Rossby left in disgrace.

His actions had been noticed, however, and he was immediately hired by the Daniel Guggenheim Fund for the Promotion of Aeronautics. His task would be to establish the first weather-reporting network for a commercial airline—Western Air Express, which flew passengers from Los Angeles to San Francisco. Its planes were not, however, the sleek jets we see today. They were awkward-looking contraptions of cloth and plywood, leftovers from World War I, without safety features or radios. Their flights along the coast of California required accurate weather information, since after the planes took off, there could be no communication with them until they (hopefully) landed safely.

The airline's current system was primitive: before they took off, pilots telephoned the airfield they were flying to and asked about the weather there. Western Air Express flights still encountered unexpected storms, since, as Rossby noted, they observed only the conditions ahead of them and "had not considered that weather may come from sideways." Rossby realized that establishing a weather-forecasting system meant gathering information from a wide variety of locations. So he traveled around California looking for people who, as he put it, "had a telephone and who stayed put all day." He often borrowed a pilot and airplane to "buzz" remote towns in areas where he had no contacts. When people came out of their homes and workplaces to watch the plane, the pilot would land and Rossby would explain what he needed. Before long, he had enlisted a group of hotel managers, gas-station attendants, and other individuals who would call Rossby's office every 90 minutes to report on visibility, winds, and precipitation in their areas.

As reports arrived by telephone, Rossby used formulas he had learned from Bjerknes to prepare weather forecasts for Western Air Express's flights. His system worked so well that it soon became the standard model for others in the young, booming airline industry. Rossby enjoyed the action. He spent evenings at nightclubs with the pilots, a crew of hard-working, hard-drinking adventurers, and usually left them feeling jealous of the way his Swedish accent seemed to captivate attractive women.

In 1928, Rossby's reputation had grown so much that the Massachusetts Institute of Technology (MIT) in Boston invited him to head its department of meteorology. In fact, MIT didn't have a department of meteorology, but it wanted to build one—and it saw Rossby as the person who could make it happen. At MIT, he was able to turn from the promotions and forecasts of the previous few years and concentrate on pure scientific research. As he gathered a devoted group of colleagues and students (something else he had learned from Bjerknes), he began to focus, not on the details of meteorology, but on the biggest of pictures. His interest turned to large-scale wind movements and the general circulation of the entire atmosphere of the planet.

At first, the MIT meteorology department consisted of just two professors, Rossby and Hurd Willett. In addition to working together, Rossby and Willett shared an apartment, and once they even dated the same woman—Harriet Marshall Alexander, who became Rossby's wife in 1929. The daughter of a Boston doctor, Alexander won Rossby's heart with her ability to identify 40 different kinds of birds by listening to their songs. The couple had two sons, Stig and Hans, and a daughter, Carin.

THE BREAKTHROUGH

The idea that there were general currents of air that spanned the globe (just as there are massive ocean currents such as the Gulf Stream) was not new. George Hadley had correctly described the general patterns in 1753. The Sun heats Earth's atmosphere unevenly, so heat builds more quickly near the Equator than it does near the poles. The warm air at the Equator rises until it reaches the top of the lowest layer of the atmosphere. When it can rise no higher, it spreads out toward the poles. As it moves north and south, it cools and sinks back toward the planet's surface. This sinking air forces more air toward the Equator, continuing the cycle. Hadley theorized that there were two of these massive cycles, one in the Northern Hemisphere and one in the Southern Hemisphere. These are often referred to as convection cells, or Hadley cells.

convection: the transfer of heat by massive motion within the atmosphere (often upward movement)

Hadley's model would work pretty well, if not for two things. First, Earth rotates, adding west-to-east movement to the smooth north-south flow described by Hadley. All free-moving objects on Earth, including winds, are deflected to the right of their paths in the Northern Hemisphere and to the left in the Southern Hemisphere. This is known as the Coriolis effect, named for Gustave-Gaspard de Coriolis (1792-1843), who first explained it in 1835. Second, the air heats up more slowly over the ocean than over the land, creating additional "ripples" in the Hadley cells.

Realizing that the existing models of atmospheric currents were much too simple, Vilhelm Bjerknes had begun to create a more accurate picture with his research into fronts in the 1910s. Bjerknes's theories about the meetings of cold and warm air masses, however, were based mainly on observations taken at ground level. He longed for more data from the upper atmosphere, but techniques to collect that data were not widely available at the time.

Rossby fared much better. During the 1930s, while he conducted his research at MIT, the United States was enduring a devastating period of drought and erosion that became known as the Dust Bowl. Anything that might help the nation predict and prepare for future disasters was a top priority. Eager to

The Dust Bowl was the greatest agricultural disaster in American history. Eager to cash in on high wheat prices, farmers had plowed and planted as much land as possible, stripping away vegetation and depleting nutrients. When drought struck in the early 1930s, the soil dried out and blew away in the strong prairie winds. Dust storms known as "black blizzards" buried entire farms, driving 3.5 million people off their land and eroding 850 million tons of soil from the Great Plains.

find ways to build better long-term forecasts, the Department of Agriculture was willing to spend money to collect data from the upper atmosphere. Rossby used this funding to further his work. At first, he and a colleague rented an airplane and flew out to make weather observations every morning. But an easier method soon emerged: radiosondes, lightweight transmitters that could be carried aloft by weather balloons. They collected data on pressure, temperature, and humidity from as high as 50,000 feet, and their movements allowed scientists to track air currents and speeds. Radiosondes were also so simple and inexpensive that it didn't matter if they were lost or damaged in flight.

Meteorologists launch a radiosonde in 1936.

When Rossby first began to gather data from radiosondes, the upper-air weather movements seemed as confused as those on the ground. As more information arrived, however, he started to see patterns. Rossby noticed massive horizontal waves that accompanied the eastward drift of the air, much as ocean waves accompany the tides. There were four or five of them, circling the entire planet in the north temperate latitudes (scientists later found them in the south temperate latitudes as well). These waves pushed and steered the movements of cold and warm air masses, which, in turn, controlled the weather. Rossby called them "long waves," but in his honor they later became known as Rossby waves. By 1939, he had gathered enough information to develop a formula that could predict the movement and speed of Rossby waves. This equation made it possible to forecast the world's weather patterns more accurately and further into the future than ever before.

In the same year as his landmark discovery, Rossby became an American citizen. He was also invited to become the assistant chief of research at the same U.S. Weather Bureau that had ousted him 12 years earlier. He accepted the position for one year, shuttling between Boston—where he continued his duties at MIT and where his wife and children lived—and Washington, D.C. In 1941, Rossby left MIT to organize a department of meteorology at the University of Chicago. (It was his philosophy that people should change their groups of friends and colleagues every 10 years or so, since by that time no one had anything new to say to anyone else.)

Rossby's discoveries about improving forecasting were desperately needed at the U.S. Weather Bureau. In 1941, its five-day forecasts were only 48 percent accurate in predicting temperatures and just 16 percent accurate for precipitation.

[Rossby] was a wonderfully stimulating teacher, an inspiring leader, and he produced ideas at a fantastic rate, but he was also a poor manager. He hardly ever answered mail. Instead, he stacked unopened letters in a pile to ripen. When they were so old that their writers no longer hoped for an answer, he felt it would do no harm to throw them away. He cut classes, was usually stony broke, ignored university budget restrictions. Sometimes he would ring furiously for his secretary when he was already dictating to her.
—*Time* magazine, 1956

This World War II poster humorously illustrated the importance of meteorology to military pilots.

Weather forecasting was crucial in the D-Day invasion of 1944, when U.S. and British forces crossed the English Channel to German-occupied Normandy, France. After storms on June 4 and 5 delayed the invasion, U.S. meteorologists foresaw a brief break in the bad weather for June 6. The attack went forward successfully, catching the Germans by surprise. They had predicted several more days of storms, making an invasion seem impossible.

He quickly gathered a respected group of scientists and students to continue his research on global weather patterns.

Then—just as it had happened with his mentor more than 20 years earlier—Rossby's work was interrupted by the immediate needs of a nation at war. World War II had begun in Europe in 1939, and the U.S. joined the conflict in 1941. The German air force's lightning-fast attacks, known as the blitzkrieg, showed that war could be fought (and possibly won) primarily by air. In response, the U.S. began building the largest air force in the world, and it needed accurate weather information to get its planes off the ground. Weather was also significant to ships and troops. A fierce storm would not only leave air forces grounded, but it would also hinder naval maneuvers and make ground fighting more difficult; a dense fog might be a hindrance for the navy while providing welcome cover for the army. Precise forecasting was a key to victory.

There were few professional meteorologists in the United States, however, and many of them were hopelessly behind the times. So the government placed Rossby in charge of a massive training program for young meteorologists. He crisscrossed the country, setting up classes at universities from coast to coast. He lectured tirelessly to groups of 400 students at a time, furiously teaching the Bergen group's methods as well as his own theories—whatever might help create better wartime forecasters.

As the war progressed, pilots of high-flying bombers began reporting changes in their air speeds.

Sometimes they traveled faster than they expected, and sometimes they made very slow progress. There were even cases reported of planes standing still in midair or actually being pushed backward by winds as strong as 200 miles per hour. Rossby determined that they were being affected by high-speed upper-air currents associated with Rossby waves. After the war ended in 1945, he developed a mathematical theory to describe these currents, which he named "jet streams" in honor of the pilots who flew with (or against) them. His equations could predict the behavior of jet steams, allowing flight planners and meteorologists to take them into account. Rossby's understanding of atmospheric currents had clearly become a sophisticated model of weather prediction.

Because air often behaves like a liquid, scientists can use water-filled tanks to study the atmosphere. The rotating movement of the tank creates waves in the water similar to those in the air. Here, Rossby poses with one of these wave tanks.

THE RESULT

In 1950, Rossby returned to the University of Stockholm, where he organized the International Meteorological Institute to promote global cooperation in meteorology. He also supported cooperation between different branches of science, finding connections between his atmospheric research and the study of Earth's oceans. Rossby knew that ocean currents could be viewed as similar to air currents, and that they could have an equally dramatic impact on the weather. He correctly theorized that Rossby waves existed in the oceans as well, traveling from east to west over long distances—even the entire span of the ocean. He noted they would be very difficult to identify, however, since their horizontal and vertical scales would be so different. An oceanic Rossby wave could have a length of hundreds of miles, while its height might be only a few inches. The waves would also move very slowly, so it could take them months or years to cross the ocean.

Scientists later verified that oceanic Rossby waves can affect the weather just as atmospheric Rossby waves do. They exert a small but constant push on currents that transport large quantities of heat through the ocean. If the Atlantic Ocean's Gulf Stream, for example, moves even slightly to the west, it affects weather patterns on the entire East Coast of the U.S., as well as changing fish and sea mammal migrations. The warm Pacific Ocean currents of 1982-1983 may have caused a Rossby wave to move across the North Pacific, eventually bringing a season of violent weather to the West Coast 10 years later.

In his final years at Stockholm, Rossby also became interested in atmospheric chemistry. The atmosphere, he noted, varied chemically—and these variations affected weather and climate. He observed that some areas of the atmosphere contained high quantities of sea salt, resulting in specific rainfall patterns. In addition, he pointed out that humans had been adding carbon dioxide (CO_2) to the atmosphere in large quantities, and that this could have drastic consequences for world climates. The next generation of meteorologists would become very familiar with the global consequences of CO_2 and other "greenhouse gases."

On August 19, 1957, at the young age of 58, Carl-Gustaf Rossby died in Stockholm. During the course of his career, meteorology had become a science of global implications. Rossby's research into large atmospheric movements had demonstrated that truly accurate prediction of the weather would require information from around the world. In the years after his death, new technologies made global forecasting a reality. Weather satellites orbiting Earth produced pictures of the entire atmosphere, providing visible evidence of Rossby waves in both the air and the oceans. And computers, using the equations that Rossby had developed, began producing weather forecasts on an even grander scale than had ever been dreamed of before.

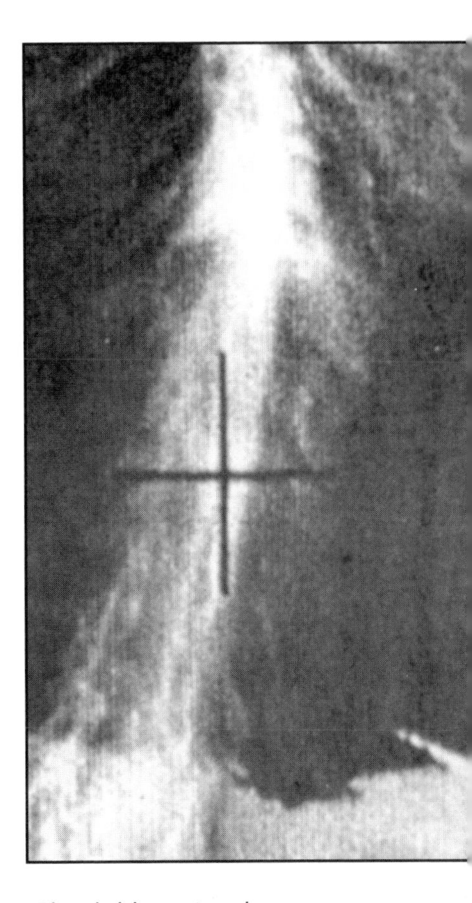

Clouds blown into long strips provide visual evidence of Rossby's jet streams. This image was captured over Africa in 1965 by the TIROS VIII *satellite. Although he did not live to see weather satellites launched, Rossby supported their development. "Right now we are like crabs on the ocean floor," he said. "What we need is a view from a satellite."*

IBM

CHAPTER SIX

Jule Charney and Computer Forecasting

At the start of the twentieth century, Vilhelm Bjerknes showed that mathematical calculations based on physical laws could be used to more accurately predict the weather. Four key elements were needed for this "numerical" forecasting. The first was data from all over the world, since weather systems move across the globe. The second was an understanding of how weather systems change—what happens when cold and warm air masses meet, for example. The third was a model that assembled all the known information to show how the atmosphere was known to behave. By the 1950s, these three pieces were mostly in place.

One problem remained, however: the fourth element needed for numerical forecasting was a way to process all the information meteorologists had gathered. The calculations were complicated by so many different variables—from air pressure to temperature to wind speed—that figuring them out took

Jule Charney (1917-1981) adapted the new technology of computers to the age-old effort of weather prediction, creating more accurate forecasts than ever before.

longer than the actual weather took to arrive. A forecast for a coming storm isn't very useful if it arrives two days after the storm has already passed. Overcoming this obstacle required the development of a technology that would change the world—and a meteorologist who knew how to use it.

On New Year's Day in 1917, Jule Gregory Charney was born to Stella and Ely Charney. Jule's parents had moved to the United States from Russia in the early 1900s and worked in the garment industry. Jule was born in San Francisco, but for most of his youth the family lived in Los Angeles. Life was difficult for Jewish immigrants, and the Charneys responded by becoming involved in socialist politics. Ely was an active union organizer, and both parents frequently discussed politics at home.

It was a lively and often intense household, and Jule responded by becoming an avid reader and a hungry learner. His bookish interests were helped along by the diagnosis of a heart problem that prevented him from enjoying the more physical joys of childhood. He often spent his free time at the public library. When he was 14, he and his mother moved to New York City for a brief time, and there, at a relative's house, Jule stumbled upon a mathematics book about calculus. The subject was not offered at any of the schools Jule had attended, so he took up the challenge of learning it on his own. To his surprise and delight, calculus helped make other sciences more understandable, and his interests began to focus on math and physics.

Jule and his mother returned to Los Angeles, and he graduated from Hollywood High School in 1934. Because family finances were limited, he attended the University of California at Los Angeles (UCLA), even though it was not known for its math and physics departments. Nevertheless, he completed college with a degree in mathematics in 1938 and with a master's in mathematics in 1940. University officials began to think that Jule Charney might become the first student in the history of UCLA to earn a Ph.D. in mathematics. But it was not going to work out quite that way.

Charney's social life expanded at UCLA. He was popular among students, who elected him King of the Mardi Gras. Medical tests had shown that the heart problem diagnosed in his childhood never actually existed, and he eagerly took to skiing and tennis.

Norwegian meteorologist Jacob Bjerknes had recently been asked to form a meteorology group within UCLA's physics department. In the spring of 1941, one of the professors in this new group invited Charney to work as his assistant and to join a meteorology training program sponsored by the army and navy. The United States would not enter World War II for another six months, but the military was already aware of the need for more scientifically trained meteorologists. If war came to the U.S., accurate weather forecasting would be essential, especially for the navy and the air force. How to make this forecasting a reality was the real challenge.

Traditional weather forecasts were based on the assumption that the atmosphere would always behave as it had before. Meteorologists studied past weather patterns in order to predict future ones. In the 1920s, however, Lewis Fry Richardson pointed out that it was unrealistic to expect complex atmospheric conditions to repeat themselves precisely.

Instead, forecasts should be based on the physical laws governing atmospheric movements. These laws could be translated into mathematical formulas that would produce exact predictions of future conditions. Richardson proposed the establishment of what he called a "forecast factory," where groups of people would calculate weather predictions in terms of numbers, and a leader would compile the results. This method, Richardson said, could forecast "faster than the weather advances." Unfortunately, it would require 64,000 people working around the clock to do so. A more efficient way was needed to make numerical forecasting a reality.

Lewis Fry Richardson (1881-1953) published his ideas about forecasting in his influential 1922 book Weather Prediction by Numerical Process, *describing the "forecast factory" as "an orchestra in which the instruments are slide-rules and calculating machines." Although he never made that vision into reality, he did test his mathematical method on his own in the 1910s. The approach proved sound, but it took him at least six weeks of work to produce a simple forecast by hand.*

After Charney joined the UCLA meteorology program as a student and teaching assistant, he was promptly set to work leading classes on synoptic meteorology—using observations of surface conditions (such as pressure, temperature, and wind) to construct weather maps. He disliked this process, which tended to emphasize elegant drawing more than objective accuracy, but at the time it was the only way for students to learn about atmospheric behavior. When Charney began his doctoral thesis in meteorology in 1944, however, he decided to take a numerical approach. He developed a number of calculations to explain how the west-to-east airflow in the middle latitudes of the atmosphere could become unstable. Even though most meteorologists were unfamiliar with such a high level of mathematics, Charney's theory was quickly published and widely accepted.

In 1946, Charney married Elinor Frye, a student at UCLA. (Elinor had a son, Nicolas, from a previous marriage, and he was soon joined by a sister and brother, Nora and Peter.) Bonds also formed in Charney's professional life that year. Visiting the University of Chicago, he met Carl-Gustaf Rossby, head of the school's department of meteorology. The two became friends, and Rossby persuaded Charney to stay in Chicago for nearly a year, discussing meteorology with Rossby, other faculty members, and distinguished visitors from overseas. Charney later described this as the most influential experience of his career.

In August 1946, Rossby arranged for Charney to attend a meeting at Princeton University's Institute for Advanced Study (IAS) about the use of computers in weather forecasting. The meeting was organized by mathematician John von Neumann, who believed meteorology was an ideal field in which to apply the newly developed electronic computers. The first of these, the Electronic Numerical Integrator, Analyzer, and Calculator (ENIAC), had been built the previous year to assist the U.S. during the final months of World War II. When Charney saw von Neumann's presentation, he realized that Princeton was quickly becoming a hub of meteorological research. He considered asking to join von Neumann's research team at IAS, and he even wrote a letter requesting a position—but never mailed it.

Awarded a National Research Council fellowship to study in Europe, Charney traveled to Norway, one of the international centers of meteorological research, in spring 1947. In his year at the University of Oslo, he refined equations to calculate atmospheric currents on a large, even planetary, scale. Then, in 1948, John von Neumann asked Charney to return to Princeton and head the meteorology group of his Electronic Computer Project.

THE BREAKTHROUGH

For the next three years, Jule and Elinor Charney lived at the IAS compound, in a wooden barracks a short walk from the building where Jule worked. Life at Princeton was exciting for the Charneys; they socialized with faculty members and met many influential and celebrated scientists of the time. Mostly, however, Jule was immersed in his research. His mission was to develop a mathematical model of the atmosphere that could be programmed into a computer von Neumann was building. Data on current atmospheric conditions would be entered into the computer, which could then use this model to calculate future air movements. Charney led a group that consisted of four other meteorologists, several computer programmers, and a secretary.

The group's first challenge was to figure out how much of the surrounding atmosphere needed to be studied in order to complete a 24-hour forecast for a given location. Using Rossby's research on atmospheric waves, Charney plotted three-dimensional west-to-east air movements and simplified them into a mathematical equation. He then added other components, such as the effect of turbulent air flowing over mountains, to create a more realistic picture of the atmosphere.

The general program developed by Charney and his group began with the different degrees of solar heating at different latitudes of the globe—in other words, the formation of warm and cold air masses. The program took into account the way these heating patterns were altered by oceans and land. Added to this global temperature map were calculations based on prevailing winds (such as jet streams) and variable winds (those affected by terrain or local weather systems), which moved air masses from place to place. The program also incorporated mathematical descriptions of what happened when air masses met—the behavior of warm and cold fronts and the formation of cyclones, as established by the research of Vilhelm and Jacob Bjerknes. All this information revealed general weather patterns, but to create forecasts that were specific enough to be useful, thousands of other factors would enter into the seemingly endless calculations.

Charney's model was ready to be tested by 1949, but von Neumann had not yet finished his computer. Instead, he contacted the army for permission to use ENIAC. The army agreed, and in April 1950, Charney's group worked around the clock to produce the first computer-based numerical weather prediction, using data gathered from 768 weather stations. Five scientists were required simply to operate the massive ENIAC, which broke down frequently in the process. The 24-hour forecast took more than 24 hours to complete, but the results were good. More importantly, the test helped the group learn how to streamline and improve its methods.

In 1952, Charney's group made its first forecast on von Neumann's computer. It began to produce forecasts of increasing complexity over the next few years, incorporating information from two different

levels of the atmosphere, then three. (One of these multilevel forecasts correctly predicted a period of intense winter storm development.) By 1953, the group was able to generate a 24-hour forecast in only 6 minutes. The project received great attention, and its numerical forecasting methods spread quickly throughout the meteorological community.

Built by John Mauchly and John Presper Eckert in 1945, the room-sized ENIAC had 18,000 vacuum tubes (which broke down about every 7 minutes), required a team of operators to work its 6,000 manual switches, and weighed 30 tons.

In his 24 years at Princeton's IAS, John von Neumann (1903-1957) built a reputation as a brilliant mathematician and computer pioneer. His work for the U.S. government included helping to develop the atomic bomb during World War II. His game theory, a mathematical decision-making method for competitive situations, is often used in political, economic, or military planning.

THE RESULT

Charney's success changed the way meteorologists prepared forecasts. While observation would continue to play an important role in predicting the weather, most forecasts would now be based on the numerical conclusions of computer programs designed to analyze tens of thousands of individual bits of data. In 1954, von Neumann and Charney worked to convince the U.S. government to use computer forecasting in the Joint Numerical Weather Prediction Unit (JNWP), a meteorological service shared by the Weather Bureau, the air force, and the navy. Charney's group taught its methods to

JNWP personnel, who began producing continuing weather forecasts in May 1955. Other sites that adopted the new numerical forecasting included the Rossby International Institute in Stockholm, Sweden, and the British Meteorological Office in Bracknell, England.

Now that numerical weather forecasting had expanded beyond Princeton, von Neumann moved on to other government assignments, and it was clear that the days of the Electronic Computer Project were numbered. Charney looked to move his research to another university, and in 1956 he and several of his research colleagues were persuaded to come to the Massachusetts Institute of Technology

By 1965, JNWP meteorologists were using an IBM 7090 computer to process weather data for forecasting, analysis, and research.

(MIT). Charney served as professor of meteorology there from 1956 until 1981.

Once Charney was established at MIT, his research moved beyond numerical forecasting to other aspects of meteorology. His interest in the movements of air and ocean currents inspired him to organize a series of informal seminars. Every two weeks, scientists from all over New England would meet to discuss their research in geophysical fluid dynamics—the study of how Earth's oceans and atmosphere move when affected by geographic features. The seminars were both social and scientific, and for 22 years they served as an informal gathering of meteorological ideas and research results.

Charney's reputation led to additional demands for his talent and his time, and the next decades held many appointments and awards. In 1957, he was appointed to the National Academy of Sciences' Commission on Meteorology. His 14-year involvement in the group gave him a higher platform to promote weather forecasting on an international scale. He particularly emphasized the use of computers, aircraft with meteorological instruments, and weather satellites. These same concerns prompted Charney to help found the National Center for Atmospheric Research (NCAR) in 1959.

In 1960, Charney was appointed to the Atmospheric Sciences Panel of President John F. Kennedy's Science Advisory Committee. He promoted the peaceful uses of space technology and the expanded role of satellites for collecting weather data. Kennedy was enthusiastic about Charney's plans and referred to them in a State of the Union address and in a 1961 speech before the United Nations. As a result, the United Nations passed resolutions that called for expanded international cooperation to improve forecasting and to recognize that weather prediction was a global concern. Charney continued to push for worldwide information-gathering in 1963, as chair of the National Research Council's Panel on International Meteorological Cooperation. In 1966, he became chair of the U.S. Committee for the Global Atmospheric Research Program (GARP), organized to map global circulation patterns in the atmosphere and use them to increase the accuracy of weather prediction.

The world's first weather satellite, TIROS I *(Television InfraRed Observation Satellite) was launched by the United States on April 1, 1960. Nine days later, it detected a typhoon in the South Pacific. Since then, many other satellites have followed, increasing scientists' knowledge of Earth, its oceans, and its weather and climate.*

chaos theory: a branch of mathematics based on the idea that small changes in equations can have large, unpredictable effects

Edward Lorenz (b. 1917) discovered the butterfly effect while running a computer weather program that rounded numbers to the sixth decimal place. After encountering a problem, he re-entered his data from a printout that only rounded to the third decimal place. Lorenz was surprised to find that the small difference of the missing decimal places led to a dramatically different result.

Meanwhile, Charney's research expanded to other related areas. In 1963, fellow MIT scientist Edward Lorenz introduced chaos theory to weather prediction with what became known as "the butterfly effect." Lorenz showed that the atmosphere's behavior was inherently unpredictable: it never returned to precisely the same state, and thus, although broad patterns might be similar, it would never precisely repeat itself. This meant that every model of the atmosphere would contain small errors that could multiply until they completely overwhelmed meteorologists' calculations. The smallest atmospheric disturbances could grow in unpredictable ways—the flapping of a butterfly's wings in China might affect storm systems the following month in New York. Charney worked with his MIT colleagues to see if they could incorporate the butterfly effect into their models. They determined that minor atmospheric fluctuations could double in approximately five days, and the outside limit for accurate weather forecasts could be about two weeks.

The freedom afforded by MIT allowed Charney to follow his scientific interests wherever they might lead. He became interested in the development of hurricanes and tornadoes, a fascination that may have dated back to 1954, when his car was damaged by a falling tree during Hurricane Carol. In 1967, Charney divorced Elinor and married Lois Swirnoff, a color theorist and professor at UCLA and Harvard. In 1972, the couple stayed for a time at the Weizmann Institute in Tel Aviv, Israel. Charney worked with scientists there to seek the meteorological causes of desertification. Several months later, the Charneys attended a summer workshop in Venice, Italy, to help reduce flooding in the city by establishing a series of floodgates. Charney's research in fluid dynamics allowed him to predict water levels in the Adriatic Sea, information necessary to the project's success.

Like his parents, Charney also followed his conscience into political activism. He opposed the Vietnam War (1954-1975) and the U.S. invasion of Cambodia (May 1970). When four student war protesters were killed by the National Guard at Kent State University on May 4, 1970, Charney worked to establish the Universities National Antiwar Fund. Over the next few months, the fund raised a quarter of a million dollars from the academic community to support antiwar candidates in the fall election.

Charney's marriage to Lois ended around 1977, and he then married Patricia Peck, a photographer. Soon, however, Charney was diagnosed with lung cancer. He died in Boston on June 16, 1981. In the course of his career, he had helped weather prediction enter the computer age, making it both more accurate and more complex. And yet Jule Charney was more than a talented mathematician and meteorologist. His view of vast weather systems allowed him to become a "global thinker." He saw the planet as he saw the weather—without local interests or national boundaries. Since this was a central aspect of his teaching, as well, the world can expect a generation of global thinkers to follow in his footsteps.

Desertification, or the spreading of deserts, occurs when loss of vegetation increases the reflection of solar radiation back into space. This reflection decreases local heating of the air near the ground. As a result, one of the primary causes of rainfall—the rising of moist, hot air into colder, upper air—does not occur. A lack of rainfall, of course, decreases vegetation further, and the cycle continues.

CHAPTER SEVEN

Howard Bluestein and Storm Chasing

On March 20, 1948, a tornado ripped through Tinker Air Force Base in Oklahoma, damaging buildings and injuring several people. Five days later, crews were still cleaning up debris when another storm began moving toward the base. On duty at the time was meteorologist and air force captain Robert Miller, who observed that the atmospheric conditions seemed similar to the ones that had led to the last tornado. Alarmed, he called commanding general Fred Borum. Borum had just one question: "Are you going to issue a tornado forecast?"

It had never been done before. Tornadoes were so brief and elusive that they were considered impossible to predict. From 1883 until 1938, the United States government had actually banned the use of the word "tornado" in weather forecasts for fear of needlessly panicking people. Even after the ban was lifted, few meteorologists thought it worthwhile to try to forecast tornadoes. Miller weighed

Howard Bluestein (b. 1948) built his adventurous scientific career on direct experience of some of the most horrible storms imaginable.

his options. The chance of two tornadoes striking the same place within five days was remote, and to warn of a tornado that didn't appear could make him seem foolish. On the other hand, what if he held back his warning and a twister did strike?

At 2:50 P.M. on March 25, 1948, Miller issued the first-ever official tornado forecast. Shortly after 6:00 that evening, a tornado hit the base. The damage was minimized, however, because people had been warned. Miller took the risk of being ridiculed and helped to prevent a possible disaster.

A large military aircraft was crushed and thrown onto parked cars by the March 25 tornado at Tinker Air Force Base. Even though Miller's tornado warning allowed base personnel to take safety precautions, the tornado still caused $6 million in damages.

The U.S. Weather Bureau finally began issuing tornado forecasts in 1952, but the warnings mainly drew criticism for their inaccuracy. Even 25 years later, meteorologists were unable to predict whether a storm would produce a tornado or where that tornado would strike. Tornadoes remained one of meteorology's last frontiers—until Howard Bluestein began his career. In his quest to understand these furious, mysterious storms, Bluestein became famous as "the Storm Chaser."

Born in Boston, Massachusetts, in 1948, Howard Bluestein was only four years old when he encountered his first tornado. Tornadoes are rare in New England, but on June 9, 1953, a twister moved through Worcester, a town just 40 miles from the Bluestein home. The storm killed 90 people, injured 1,200, and caused $52 million in property damage. Severe weather seemed to play a recurring role in Howard's early life. "When I was very young," he noted, "lightning struck a TV antenna on our house and our TV blew up in my face. That got my attention! We also had Hurricane Carol, which passed over in 1954. I remember going out when the center of that storm passed by. As I remember it, the eye passed overhead . . . and the rain stopped and the wind stopped, and people poured out of their houses to look at the damage. And then I remember the winds shifted to the other direction and got pretty strong, and I had to fight my way back home."

It's easy to see how young Howard's interests turned to the weather, but storm chasing was not an early ambition. His first weather-related hobby was photography. After his parents gave him both a telescope and a camera, Howard's first photos were of the sky. "It's nice to have parents that take an interest in what you do and allow you to go where your interests lead," he later recalled. "I had parents who, when I was interested in electronics, got me a crystal radio set, and when I was interested in photography they got me a camera." Later, Howard focused on photos of cloud formations. His experiments with photography would play an integral role in studying tornadoes and conveying his discoveries. (His 1999 book, *Tornado Alley: Monster Storms of the Great Plains*, was illustrated with his own photographs.)

Another key childhood interest was electronics. In fact, electrical engineering was Bluestein's chosen major as an undergraduate student at the Massachusetts Institute of Technology (MIT). He earned his bachelor of science degree in 1971 and his master's in 1972, also in electrical engineering. But during a summer job at MIT working for professor of meteorology Fred Sanders, he became hooked on the weather. In 1976, Bluestein received his Ph.D. in meteorology. At about that time, he met Ed Kessler, the founder and director of the National Severe Storms Laboratory (NSSL) in Norman, Oklahoma. Kessler told exciting stories of the storms in Oklahoma and asked Bluestein if he would like to move there to study tornadoes. Bluestein couldn't pass up the opportunity to work in the middle of "Tornado Alley," the most tornado-prone area in the world. He got a job teaching at Oklahoma University (also in Norman), and the Storm Chaser was born.

Tornadoes are one of nature's most spectacular displays, with more destructive power than any other kind of violent storm. Even though predicting if and where actual tornadoes will strike is difficult, the general conditions for the formation of a tornado are well known. Tornadoes are typically a product of the severe thunderstorms that form when warm, moist air near the ground underlies cooler, drier air moving in at a higher altitude. The warmer air rises over the cooler air, bringing moisture into higher elevations, where it condenses into clouds. If the temperature differences are great enough, then the moist air rises until it reaches the top of the troposphere, the lowest layer of Earth's atmosphere. There it can rise no further, so it spreads out into anvil-shaped thunderclouds. This violent updraft of rising air is accompanied by an opposite downdraft, as rain forms and begins to fall. If there is enough of a difference between the wind speeds on the ground and those in the upper air, the cloud system can begin to rotate. A funnel-shaped cloud may form, extend downward, and touch the ground. This spinning column of air beneath a thunderstorm is a tornado.

The word "tornado" probably comes from the Spanish word *tornar,* meaning "to turn."

An anvil thundercloud

Although no one can prevent or control these devastating storms, the key to coexisting with them is knowing when they might appear and then taking steps to minimize the damage and injury they can cause. When Bluestein began his storm-chasing career, however, few observations of tornadoes were made by scientists. The existing photographs and films were mostly captured by people who happened to be caught in storms. Bluestein and his Oklahoma

The United States is the "tornado capital" of the world. Typically, about 700 twisters occur in the U.S. each year, far more than in any other country. They most frequently appear in "Tornado Alley," a 100-mile-wide area including northern Texas, Oklahoma, Kansas, and Nebraska. But tornadoes can strike anywhere, and the effects can be devastating. On April 11, 1965, a chain of up to 50 tornadoes killed 256 people in 6 states; on April 3-4, 1974, 148 tornadoes killed 300 people in 13 states.

During these early days of chasing, the public knew virtually nothing of our activities. People often spotted us driving erratically, stopping, starting, moving quickly, or creeping along. Complaints to the university were frequent, and we were often warned to be more discreet.
—Howard Bluestein

colleagues were the first to suggest that meteorologists might benefit from systematically seeking out and observing as many tornadoes as possible. Beginning in 1977, Bluestein helped to organize storm-chasing teams of graduate students—and, occasionally, lucky undergraduates—from Oklahoma University's School of Meteorology (often working with the NSSL). Their goal was to locate potential tornadoes, drive as near to them as they could, and film or photograph the storms. Their observations of conditions on the ground would be compared to radar readings to help determine what signs forecasters should look for when trying to predict tornadoes.

Every spring, Bluestein and his teams logged between 5,000 and 10,000 miles driving across Oklahoma, Texas, New Mexico, and Kansas in their quest for tornadoes. The team's efforts, however, were not always successful. Tornadoes most often occur in the Great Plains of the midwestern United States from April through early June, but they can happen anytime and anywhere. Even under the best storm conditions, catching a tornado is tricky business. As Bluestein noted, "Tornadoes don't last very long, and their paths change, and the road system doesn't allow you to get to every location. We would go out some days, and the tornado would be coming toward us . . . and the tornado would either disappear before getting to us or change course. If we think it's going to happen, we go," he added. "Then we will try to get ourselves in the path, usually to within a few miles." Despite their best efforts, Bluestein estimated, his storm-chasing teams saw a tornado only about once out of every nine trips out.

By the 1980s, funding for Bluestein's work was becoming scarce. "Some within the meteorology community looked upon storm chasers as merely thrill seekers, engaged in a dangerous sport," he said. "Storm chasing needed a better image." Bluestein believed that storm chasing could improve its image—and its results—if teams could make quantitative measurements of conditions in and near the tornadoes. Drawing on his background in engineering, he began seeking instruments that would help him "see inside" a tornado.

Bluestein (standing next to truck) chased down this tornado in Happy, Texas.

THE BREAKTHROUGH

Decommissioned in 1986, TOTO is now on display at the NOAA headquarters in Washington, D.C.

In 1980, two scientists from the National Oceanic and Atmospheric Administration (NOAA), Al Bedard and Carl Ramzy, approached Bluestein with an idea for building a portable observatory to study tornadoes. The result was the Totable Tornado Observatory, nicknamed TOTO (partly in remembrance of Dorothy's dog in *The Wizard of Oz*). The device contained several basic meteorological instruments to measure such things as air pressure, wind speed and direction, and temperature, all housed in a 400-pound case designed to withstand direct contact with a tornado. The plan was to carry TOTO in the back of a pickup truck, maneuver into the path of a tornado, roll TOTO down the truck's ramp, turn on the instruments, and head for safety. With luck, the tornado would pass over TOTO, which would then be able to collect information about the interior of the storm. Unfortunately, TOTO never provided the hoped-for data. Getting the device into the precise path of a tornado at just the right time proved difficult and even dangerous. It began to seem, Bluestein later wrote, "as though tornadoes tended to avoid TOTO." By 1983, he had concluded that there had to be a better way.

The solution arrived in the form of Doppler radar. Radar (an acronym for Radio Detection And Ranging) had been used by meteorologists since World War II. Radio waves were transmitted from a source point through the atmosphere, and when they struck an object—whether it was an airplane or a raincloud—they bounced off of it and returned to the source. Radar could pinpoint the location of the object and, based on the amount of reflected energy, its size. In the 1980s, the National Weather Service began to invest in an improved form of radar that could also indicate the speed and direction in which an object was moving. This radar analyzed the Doppler effect from a signal, or the change in frequency (number of waves per unit of time) caused by an object moving toward or away from the observer. If rain was being carried toward the signal's source, the frequency of the reflected radio waves would be higher; if the rain was moving away, the frequency would be lower. Doppler radar could show not only the movements of a storm system as a whole, but also the speed and direction of winds within the system. It could even detect winds in cloudless air.

The Doppler effect is named after Christian Johann Doppler (1803-1853), who first identified it. A common example of the Doppler effect is the whistle of a moving train. If the train is moving toward someone standing by the side of the tracks, that person will hear the pitch of the whistle becoming progressively higher as the frequency of the waves increases. When the train moves away from the observer, the whistle's pitch will sound lower as the waves spread out.

Around the time of Bluestein's frustrations with TOTO, Doppler radar had progressed to the point where the radar device could be made mobile. Bluestein saw the potential of this new technology, and in 1987 he was able to collect data from a tornado without actually placing a device inside it. Instead, his storm-chasing teams could observe tornadoes from several miles away. By revealing the speed and direction of the swirling winds within a tornado, portable Doppler proved the best—and safest—method to better understand how these storms behaved.

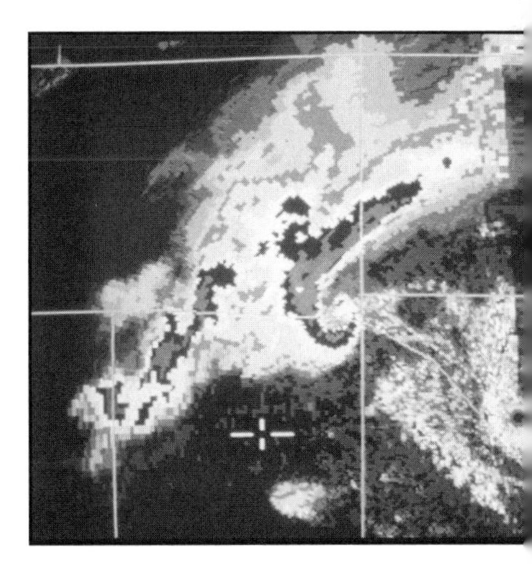

A Doppler radar image showing a "hook echo," a sign of a rotating cylinder of air that indicates possible tornado formation

Bluestein could not always stay comfortably distant from the twisters he studied, however. In 1991, he came within a mile of the path of a fierce

Bluestein's crew members—graduate students Doug Speheger (left), Jim LaDue (top), and Herb Stein—use mobile Doppler to probe a tornado near Enid, Oklahoma, on April 12, 1991. The radar instrument is housed in the rectangular box mounted on the tripod in front of LaDue, with two dish antennas to transmit and receive the signals. The chest on the ground contains batteries and recording equipment.

storm in Red Rock, Oklahoma. "This huge tornado blew a house right off its foundation," he later recalled. "The tornado leaped across the road right in front of us. You could see its furious spin. It was very, very clear." When Bluestein used a mobile Doppler system to measure the storm's wind speed, it clocked in at 286 miles per hour—the highest ever recorded on Earth. He thus became the first person to officially measure an F5 category tornado.

Bluestein had another "close encounter" in May 1999, when he was able to bring a mobile Doppler unit within a half-mile of a tornado's funnel. He recorded the process of the tornado's formation, including the patterns of dust and rain as the twister developed, and was able to create a clear depiction of the wind patterns inside the storm. The data he gathered would help build a better understanding of a tornado's structure and life cycle.

Bluestein's photograph of the F5 tornado. The force of a tornado is expressed on the Fujita scale, developed by Tetsuya "Ted" Fujita in 1971. The scale categorizes storms by their wind speed, ranging from F0 (winds between 40 and 72 miles per hour) to F5 (greater than 260 mph).

THE RESULT

Bluestein's storm chasing helped bring attention to tornado research, both within the scientific community and among members of the public. Mobile Doppler radar became a standard tool for other researchers, yielding data that continually brought them closer to understanding the dynamics of a tornado. The observations collected by Bluestein and other storm chasers were the basis for the tornado warning procedures instituted by the National Weather Service in the early 1990s. Thanks to their identification of the conditions that tend to cause tornadoes, the time between when a warning is issued and when a tornado actually strikes has increased by as much as 30 minutes.

The 1996 movie *Twister* was based in part on Bluestein's work (it portrays a group of storm chasers who use a TOTO-like device called "Dorothy"). Although the film increased popular interest in storm chasing, it was not scientifically accurate; Bluestein called it "a comic-book version of what meteorologists do."

Despite these triumphs, however, Bluestein emphasized that much remained to be learned about tornadoes. Scientists still have yet to understand why some thunderstorms produce tornadoes and others do not. Thus, forecasting remains far from accurate. "You need to understand how the tornado forms," said Bluestein. "Right now, the National Weather Service looks for rotation in the storms . . . and issues a Tornado Warning. But [then] you issue a lot of false warnings. People don't get very excited if you end up 'crying wolf.'" A key problem, according to him, is that "There isn't usually just one thunderstorm. Thunderstorms interact with each other. One of the things I'm working on right now is the study of storm interaction to see how that affects the behavior of storms."

Having seen more than 100 tornadoes in his first 25 years as a storm chaser, Bluestein plans to continue his work for as long as possible. "I still see things each year that I have never, ever seen before," he said in 2002. "I'm still fascinated." His new projects include analyzing data from mobile Doppler units and other instruments, adding data compiled by the National Weather Service, and developing sophisticated computer models that could highlight other areas for future investigation. "With new radar technology and faster and larger computers," he told a committee of the U.S. Congress in June 1999, "we can make great progress in the next five to ten years in determining how and why tornadoes form, what new instruments will be required to make the necessary measurements to do so, and how to protect the citizenry."

And so every year, as spring rolls around, Howard Bluestein wraps up his classes at Oklahoma University, organizes his gear and his crew, and once more takes to the roads in search of tornadoes. "In order to study a meteorological phenomenon properly," he believes, "you must actually experience it and appreciate it aesthetically." Yet no matter how many mysteries are unraveled about tornadoes, they will remain as frightening and fascinating as ever. "Our quest for discovery has not taken away from the respect we have for the awesome power the tornado harbors, nor the thrill of viewing the violent motions in the tornado or the beauty of the storm," Bluestein has written. "We eagerly await the next act in the atmospheric play starring the tornado."

Bluestein sometimes went to great lengths to catch a storm. In 1992, he accomplished what he called "the first and only . . . hybrid bicycle-car chase," frantically pedaling home to borrow his wife's car in order to follow a tornado. Earlier in his storm-chasing career, while working on the fourth floor of the National Hurricane Research Laboratory in Florida, Bluestein became so fascinated by a nearby storm that, he remembered, one of his colleagues "had to yank me back, when, exhilarated, I teetered precariously close to the edge of the ledge while photographing a funnel cloud."

When asked whether he thought tornadoes were beautiful, Bluestein did not hesitate to reply. "Sure—as long as they're not killing anybody. When they're out over open country, they're beautiful. When they're moving through a city, they're scary."

GOES-1 DPT 298 1645Z 25 OCT. 75

AFTERWORD

Extending the Limits

In 1891, scientist John Pillsbury offered the following vision of the future of meteorology: "we will, with the aid of meteorology, be able to say to the farmer hundreds of miles distant from the sea, 'You will have an abnormal amount of rain during the next summer,' or 'The winter will be cold and clear,' and by these predictions they can plant a crop to suit the circumstances or provide an unusual amount of food for their stock." More than a century later, however, meteorologists still seem short of this goal. Has science reached the limits of reliable weather prediction? Where does meteorology go from here?

Vilhelm Bjerknes's mathematical approach to forecasting has been refined and new variables added so that meteorologists can now create 10-day forecasts with some sense of reliability. Still, the daily forecasts most people see on their local television news programs typically extend only 5 days into the future. Reliability drops significantly when a 5-day

Since October 1975, when this first image was taken, Earth has been observed by a series of satellites known as GOES (Geostationary Operational Environmental Satellites). "Geostationary" means that they remain over the same fixed point, keeping pace with Earth's rotation from 22,000 miles above. Between them, two GOESes provide images of the entire planet at once.

forecast is extended to 10 days, and TV meteorologists are reluctant to place their credibility in such danger. Will scientists ever be able to provide accurate season-long forecasts?

In a sense, they already do. In the 1970s, scientists James Lovelock and Lynn Margulis presented their Gaia Theory, named after the Greek goddess of Earth. The theory presented a view of the planet as a single, self-regulating organism in which every individual species would act to ensure that the environment remained hospitable for life to survive. Thus, minor changes to the planet would be countered naturally. Greater changes in the ecosystem, however, might extend beyond Earth's ability to correct. In particular, human-created pollution and land development could have devastating effects on the environment. The Gaia Theory made scientists more aware of the many interrelated factors that affect the entire planet. Meteorologists now examine much more information in an effort to create better long-range weather predictions—including the current amount of sea ice, the wobbling of Earth about its axis, and the level of solar activity. Benjamin Franklin would be glad to know that his early concept of the greenhouse effect has been factored into weather forecasts, too.

There are two jet streams in each hemisphere, the polar and the subtropical. When U.S. meteorologists refer to "the jet stream," they mean the Northern Hemisphere's polar jet stream, which marks the boundary between cold Canadian air and the warmer air to the south. When this current shifts, it can cause sudden changes in the weather. On December 24, 1989, the polar jet dipped south, plunging temperatures in Miami to 33 degrees Fahrenheit and destroying the Florida citrus crop.

The effect of large movements of the atmosphere and ocean, as established by Carl-Gustaf Rossby, is also emphasized in modern weather prediction. Scientists routinely map jet streams, and TV weather forecasters often explain the effect of these upper-air currents to their viewers. Ocean currents also make the news frequently. Most of North America has heard of the periodic warming of Pacific Ocean currents known as El Niño, and meteorologists frequently discuss how this phenomenon might affect weather for the season ahead.

The use of computers has moved far beyond even the dreams of Jule Charney, becoming so central to weather prediction that meteorologists often run the same atmospheric model 10 to 40 times, each time incorporating a minor change—just to see how that change might affect the long-range picture. These are referred to as "ensemble forecasts,"

This image of the Pacific Ocean taken by satellite in 1997 shows El Niño as an area of warm surface water (in white, on right) about one and a half times the size of the continental United States. The El Niño phenomenon is triggered every four to seven years when west-blowing trade winds weaken, allowing a mass of warm water normally located near Australia to move eastward along the Equator toward South America. This affects evaporation and the formation of rain clouds, changing jet stream patterns—and therefore weather—around the world.

There have been improvements not only in making forecasts, but also in communicating them to the public. Perhaps the most dramatic came in 1982, when TV weathercaster John Coleman launched The Weather Channel (TWC) to broadcast local, national, and international weather 24 hours a day. Below, TWC staff members are briefed by computer on the latest conditions.

since several variations appear simultaneously. If the ensemble shows little variation, then meteorologists can assume conditions are relatively stable. On the other hand, if minor changes cause major variations in a 10-day forecast, then conditions are unstable and meteorologists are wary of making any substantial predictions. Ensemble forecasts often indicate the probability of an event, rather than simply saying the event will happen—for example, "a 40 percent chance of rain."

The organizations that gather data about the weather have also grown. In the United States, the

National Oceanic and Atmospheric Administration (NOAA) compiles information about the oceans and the atmosphere. It is the parent organization of the National Weather Service (NWS), which serves as the authority for all weather matters in the U.S., with the responsibility of issuing watches and warnings for severe storms, hurricanes, and tornadoes. National weather services operate in most other countries, and they are becoming part of a global chain of meteorological services. For example, the European Center for Medium-Range Weather Forecasts in England offers forecasts for 17 member countries. The World Meteorological Organization (WMO), a branch of the United Nations, coordinates research and forecasting for 185 member countries.

And things are looking up. Way up, in fact. Meteorology has become a principal focus of the U.S. and international space programs. Satellites can capture images of cloud cover, temperature patterns, ocean currents, and even volcanoes and forest fires. In the 1980s, the National Aeronautics and Space Administration (NASA) began its Earth Observing System (EOS) to collect long-term observations of the global environment and climate. The EOS flagship satellite, *Terra*, was launched in 1999, and another satellite, *Aqua*, followed in 2002. *Aqua* contained six instruments that would gather information about water on Earth—including oceans, humidity, and precipitation—to compile a continuous record of the world's hydrologic cycle. NASA estimated that data gathered by *Aqua* could increase the useful range of weather forecasts by two or three days.

A computer-generated model of the Aqua *satellite*

So how far will such advances take us in the quest for better and longer forecasting? While there is no clear answer, it might be best to consider a weather forecast as being one of two different types. The daily forecast of what is going to happen for a local area (When is it going to rain? How hot will it be today?) may not move much past a 5-to-10-day limit, despite additional data. The public demands accurate details in daily forecasts, and there are many minor variables that can affect the weather. In a period of perhaps a week, these variables can grow until they alter the details of even the best forecasts.

On the other hand, forecasts of general trends may extend six months in advance. Meteorologists may, in fact, be able to tell farmers that the next summer will be drier than usual, or the next winter colder—and do it with a good deal of detail. It isn't difficult to see the value of such long-range predictions. Every once in a while, for example, the U.S. sees a summer so hot and dry (especially in the West) that wildfires ignite into life-and-death battles that last for weeks over tens of thousands of acres. A prolonged season of rain in the Midwest can cause disastrous flooding along the Mississippi and other rivers. Someday soon, the public may receive effective warnings about such potential dangers and be able to take steps to minimize their harm to wildlife, property, and people. These general forecasts won't predict what will happen on a particular day, but they are an indispensable part of weather awareness and preparation.

An image from the Moderate Resolution Imaging Spectroradiometer (MODIS), one of the six instruments aboard Aqua. *Taken on September 22, 2002, it shows Hurricane Isidore over the Yucatán Peninsula. Several people were killed and more than 300,000 left homeless after 125-mile-per-hour winds and flooding rains struck Mexico.*

And hasn't that been a goal of meteorology from the start? Preparing a farmer for a harsh winter serves the same general purpose as asking people to place lightning rods on their homes, or trying to figure out whether a particular storm might produce a tornado. Meteorologists might never be able to help us control the weather. Nevertheless, understanding the weather and reliably predicting how it might affect us are more realistic goals. And modern meteorology is moving toward these goals in more areas than ever before.

GLOSSARY

air mass: a large body of air within which the temperature, pressure, and humidity are constant at all altitudes

astrology: the study of the positions of the Sun, Moon, stars, and planets in the belief that they influence earthly events and human fortunes

atmosphere: the gaseous mass surrounding Earth or any other planet

atmospheric pressure: pressure caused by the weight of the atmosphere (also called barometric pressure)

barometer: a device used to measure atmospheric pressure

calibrate: to set, check, or correct the graduations of a measuring instrument

Celsius scale: originally, a scale of temperature measurement in which 0° was the boiling point of water and 100° was the freezing point; now used to describe a scale in which 0° is the freezing point and 100° is the boiling point

Centigrade scale: a scale of temperature measurement in which 0° is the freezing point of water and 100° is the boiling point; now commonly called the **Celsius scale**

chaos theory: a branch of mathematics based on the idea that small changes in equations can have large, unpredictable effects

cirrus: a high, thin cloud

condensation: a substance's change of state from a gas to a liquid

convection: the transfer of heat by massive motion within the atmosphere (often upward movement)

Coriolis effect: the effect of Earth's rotation on all free-moving objects, including the atmosphere and oceans. Objects are deflected to the right of their paths in the Northern Hemisphere and to the left in the Southern Hemisphere.

cumulus: a dense, white, fluffy cloud with a flat base and a rounded top

cyclone: an area of low atmospheric pressure characterized by rotating and converging winds and rising air, usually accompanied by stormy weather

Doppler effect: the change in frequency caused by the movement of an object relative to an observer

Doppler radar: radar that uses the Doppler effect to detect the speed and direction of an object in motion

electrodynamics: the branch of physics that studies electric charges and currents

electromagnetic waves: waves made up of an electric field and a magnetic field that have the same frequency and travel at the speed of light (kinds of electromagnetic waves include radio waves, X-rays, and ultraviolet rays)

El Niño: a periodic warming of Pacific Ocean currents that affects world weather

evaporation: a substance's change of state from a liquid to a gas

Fahrenheit scale: a scale of temperature measurement in which 32° is the freezing point of water and 212° is the boiling point

front: the boundary between two air masses of different temperatures and densities. In a **cold front**, cold air moves in on warm air; in a **warm front**, warm air moves in on cold air; in a **stationary front**, the air masses do not move; and in an **occluded front**, a cold front overtakes a warm front. The **polar front** is the stormy boundary separating cold polar air masses from warm tropical ones.

Gulf Stream: a warm Atlantic Ocean current that flows north from the Gulf of Mexico

Hadley cells: two systems of circulation (one in each hemisphere) that arise from uneven heating of the atmosphere, which causes warm air to rise and move away from the Equator and cold air to sink and move toward it; also called convection cells

humidity: the amount of water vapor in the air

hydrodynamics: the branch of physics that studies the motion of fluids

hydrologic cycle: the continuous circulation of water through the biosphere

jet stream: a narrow band of fast-moving, high-altitude air that flows along the boundary between warm and cold air masses

Kelvin scale: a scale of temperature measurement in which 273.16 is the point that water, water vapor, and ice can coexist

Leyden jar: a device used to contain an electrical charge

lightning rod: a metal rod placed atop a building or ship to conduct lightning safely to the ground

mercury: a heavy, silver-white metallic element

meteorology: the study of weather, including weather forecasting

nephology: the study of clouds

nomenclature: a system of naming

precipitation: any form of water, such as rain or snow, that falls to the earth's surface

radar: a method of detecting distant objects by analysis of radio waves reflected from their surfaces

radiosonde: a lightweight package of weather instruments with a radio transmitter, carried aloft by a balloon

Rossby waves: long upper-air waves in the middle latitudes (also found in oceans)

satellite: an object launched to orbit Earth or another celestial body, often used to observe weather

stratus: a low cloud forming in large sheets or layers

thermodynamics: the branch of physics that studies the transformation of heat to other forms of energy

thermometer: an instrument that measures temperature quantitatively

tornado: a rotating column of air that forms beneath a thunderstorm

BIBLIOGRAPHY

"About Temperature." www.unidata.ucar.edu/staff/blynds/tmp.html.

Ahrens, C. Donald. *Meteorology Today: An Introduction to Weather, Climate, and the Environment.* St. Paul: West Publishing, 1988.

Asimov, Isaac. *Asimov's Biographical Encyclopedia of Science and Technology.* New York: Doubleday, 1982.

Barnes-Svarney, Patricia, ed. *The New York Public Library Science Desk Reference.* New York: Simon & Schuster, 1995.

Bluestein, Howard B. Telephone interview. April 6, 2002.

———. *Tornado Alley: Monster Storms of the Great Plains.* New York: Oxford University Press, 1999.

Bluestein, Howard B., and Andrew L. Pagmany. "Observations of Tornadoes and Other Convective Phenomena With a Mobile, 3-Man Wavelength Doppler Radar: The Spring 1999 Field Experiment." *Bulletin of the American Meteorological Society,* Vol. 81, No. 12.

Bluestein, Howard B., et. al. "Meeting Summary: Ground-Based Mobile Instrument Workshop Summary, 23-24 February 2000, Boulder, Colorado." *Bulletin of the American Meteorological Society,* Vol. 82, No. 4.

Bodin, Svante. *Weather and Climate.* Dorset: Blandford Press, 1978.

Bowen, Catherine D. *The Most Dangerous Man in America.* Boston: Little, Brown, 1974.

Brands, H. W. *The First American: The Life and Times of Benjamin Franklin.* New York: Doubleday, 2000.

Burke, James. "Cool Stuff." *Scientific American,* July 1997.

Burroughs, William J., Bob Crowder, Ted Robertson, Eleanor Vallier-Talbot, and Richard Whitaker. *Weather.* New York: Time-Life, 1996.

Cantrell, Mark. *The Everything Weather Book.* Avon, Mass.: Adams Media, 2002.

Eliassen, Arnt. "Jacob Aall Bonnevie Bjerknes: Biographical Memoirs." www.nap.edu/readingroom/books/biomems/jbjerknes.html.

European Geophysical Society. "Vilhelm Bjerknes." www.copernicus.org/EGS/egs_infor/bjerknes.htm.

Fleming, Thomas. *The Man Who Dared the Lightning: A New Look at Benjamin Franklin.* New York: Morrow, 1971.

Friedman, Robert Marc. *Appropriating the Weather: Vilhelm Bjerknes and the Construction of a Modern Meteorology.* Ithaca, N.Y.: Cornell University Press, 1989.

Gillispie, Charles Coulston, ed. *Dictionary of Scientific Biography.* New York: Scribner's, 1972.

Hamblyn, Richard. *The Invention of Clouds: How an Amateur Meteorologist Forged the Language of the Skies.* New York: Farrar, 2001.

Heidorn, Keith C. "Luke Howard: The Man Who Named the Clouds." www.islandnet.com/~see/weather/history/howard.htm.

Heninger, S. K., Jr. *A Handbook of Renaissance Meteorology.* Durham, N.C.: Duke University Press, 1960.

Laskin, David. *Braving the Elements: The Stormy History of American Weather.* New York: Doubleday, 1996.

Lutgens, Frederick K., and Edward F. Tarbuck. *The Atmosphere: An Introduction to Meteorology.* Sixth edition. Englewood Cliffs, N.J.: Prentice-Hall, 1995.

Lynch, John. *The Weather.* Toronto: Firefly, 2002.

"Man's Milieu." *Time,* December 17, 1956.

Middleton, W. E. Knowles. *A History of the Thermometer and Its Use in Meteorology.* Baltimore: The Johns Hopkins Press, 1966.

National Oceanic and Atmospheric Administration. www.noaa.gov.

NOVA Online. "A Day In the Life of a Stormchaser." www.pbs.org./wgbh/imax/life.html.

Phillips, Norman A. "Jule Gregory Charney: Biographical Memoirs." www.nap.edu/readingroom/books/biomems/jcharney.html.

QPB Science Encyclopedia. New York: Helicon, 1998.

Schneider, Stephen H., and Londer, Randi. *The Coevolution of Climate and Life.* San Francisco: Sierra Club, 1984.

Shreve, Jenn. "Storm Chaser: Tornado Expert Howard Bluestein Says that Cows Don't Fly, but Cars Do." *Salon,* July 19, 1999. www.salon.com.

Southampton Oceanography Centre. "Rossby Waves: What Are They?" www.soc.soton.ac.uk/JRD/SAT/Rossby/Rossbyintro.html.

Van Doren, Carl. *Benjamin Franklin.* New York: Viking, 1938.

"Vilhelm Frimann Koren Bjerknes." www-gap.dcs.st-and.ac.uk/~history/Mathematicians/Bjerknes_Vilhelm.html.

The Weather Channel. www.weather.com.

INDEX

ABOUT THE AUTHORS

Susan and Steven Wills, both educators and writers, are the authors of *Astronomy: Looking at the Stars* from The Oliver Press. Steven, a teacher for the last 33 years, is also the author of *Mind-Boggling Astronomy* and writes frequently for science magazines. His weekly newspaper education column has been syndicated since 1984. Susan, with over 20 years of classroom experience, has written science articles for more than a dozen children's and young adult magazines. Susan and Steven live and work near Philadelphia, Pennsylvania.

PHOTO ACKNOWLEDGMENTS

Margaret Barnes: p. 144
Howard Bluestein: pp. 110 (Andrea Wilson), 120, 121
Charles Babbage Institute, University of Minnesota: p. 103
Gimnazjum 25, Gdansk, Poland: p. 16
Hulton/Archive by Getty Images: pp. 27, 65
Matthew R. Kramar: p. 117
Library of Congress: pp. 12, 15, 30, 33, 34, 37, 38, 39, 45, 54, 104, back cover
Mary Evans Picture Library: pp. 28, 52
MIT Museum: pp. 94, 108
Museum Boerhaave Leiden: p. 25
NASA: pp. 127 (TOPEX/Poseidon, NASA JPL), 130 (TRW), 131 (Jacques Descloitres, MODIS Land Rapid Response Team, NASA/GSFC)
National Archives: p. 90
National Library of Medicine: p. 55
National Meteorological Library: p. 98 (copyright Elliott & Fry)
NOAA: cover, pp. 6, 8, 9, 19, 42, 44, 75, 77, 78, 79, 83, 87, 88, 91, 93, 105, 107, 112, 115, 118, 119, 124
Revilo: pp. 20 (both), 26
Royal Meteorological Society: p. 46
Science Museum/Science and Society Picture Library: pp. 56, 57, 60
University of Chicago Library: p. 80
University Library of Bergen, Picture Collection: pp. 62, 72, 74
The Weather Channel, Inc.: p. 128
Yerkes Observatory: p. 21